Stefan Krebs, Heike Weber (eds.)
The Persistence of Technology

Science Studies

Stefan Krebs (Dr. phil.) is Assistant Professor for Contemporary History at the Luxembourg Centre for Contemporary and Digital History.
Heike Weber (Dr. phil.) is Professor of History of Technology at the Technische Universität Berlin.

Stefan Krebs, Heike Weber (eds.)

The Persistence of Technology

Histories of Repair, Reuse and Disposal

[transcript]

Supported by the Luxembourg National Research Fund (FNR) (Project ID: 12547405) and the Luxembourg Centre for Contemporary and Digital History (C²DH).

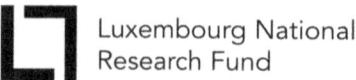

Bibliographic information published by the Deutsche Nationalbibliothek
The Deutsche Nationalbibliothek lists this publication in the Deutsche National-bibliografie; detailed bibliographic data are available in the Internet at http://dnb.d-nb.de

First published in 2021 by transcript Verlag, Bielefeld
© Stefan Krebs, Heike Weber (eds.)

Cover layout: Maria Arndt, Bielefeld
Cover illustration: seraph / photocase.com
Proofread by Sarah Cooper, Miriam Verena Fleck
Typeset by Stefan Krebs

Print-ISBN 978-3-8376-4741-9
PDF-ISBN 978-3-8394-4741-3
https://doi.org/10.14361/9783839447413
ISSN of series: 2703-1543
eISSN of series: 2703-1551

Contents

Preface

Stefan Krebs and Heike Weber

The papers in this volume were first discussed during a workshop at the Luxembourg Centre for Contemporary and Digital History (C²DH) in December 2018. We would like to thank all the workshop participants for their contributions. In particular, we would like to thank Patrick Fridenson for his final comments. He neatly summarised the papers and wrapped up the discussions, and he also pointed out the gaps in the historiography of repair that we could not cover during the workshop and that we are also unfortunately unable to address in this volume. He pointed out, for example, that the role of technical education for professionals and consumers should be taken into account when studying maintenance and repair cultures. The same goes for health and safety as well as environmental regulations that have shaped maintenance and repair practices. Furthermore, he reminded us that despite the vast geographical scope of the workshop contributions, Africa was completely absent from our considerations – although we know from ethnographic studies that African repair cultures are important in understanding the persistence of technology in the Global South.

We would like to express our gratitude to our copy editors Sarah Cooper and Miriam Verena Fleck for doing a fantastic job finalising this anthology, and to Rebecca Mossop and Thomas Hoppenheit, PhD students in the "Repairing technology – fixing society? History of maintenance and repair in Luxembourg" (REPAIR) project, for a critical reading of the final manuscript. In addition, we would like to thank the Luxembourg National Research Fund (FNR) for supporting the publication of this volume as part of the CORE project C18/SC/12547405.

This collection was originally due to be published in spring 2020 and the authors' chapters reflect the state of research at that time. The delay is the sole responsibility of the editors.

The Persistence of Technology:

From Maintenance and Repair

to Reuse and Disposal

Heike Weber and Stefan Krebs

Today, the act of repair has developed into a kind of social movement. The repairers who meet in repair cafés and other such venues, assisted by various organisations, forums and online platforms, are driven by the political idea that by fixing objects, they can fix the world and its predominantly capitalist economic model.[1] In their view, repairing is associated with sustainability goals; it is seen as an act of environmentalism.[2]

Yet at its core, repairing is, and will remain, a user operation on objects and goods; it is a fundamental interaction between humans and technology. According to Stephen Graham and Nigel Thrift, repair and maintenance constitute "the engine room of modern economies and societies".[3] Henke and Sims see repair at work in any process which restores social or material order, but they particularly emphasise the role of infrastructure repair in today's interconnected, standardised world – interventions that encompass local fixes as much as systemic approaches or efforts to "reflexively" repair the unintended environmental consequences of modern infrastructures.[4] This omnipresence makes it somewhat

---

1   See https://repaircafe.org/en/foundation, ifixit.com or www.reparatur-initiativen.de (all accessed 04.11.2019).

2   See Baier, Andrea et al. (eds.): Die Welt reparieren: Open Source und Selbermachen als postkapitalistische Praxis, Bielefeld: transcript 2016.

3   Graham, Stephen/Thrift, Nigel: "Out of Order: Understanding Repair and Maintenance", in: Theory, Culture & Society 24, 3 (2007), p. 1–25, here p. 19.

4   Henke, Christopher R./Sims, Benjamin: Repairing Infrastructures: The Maintenance of Materiality and Power, Cambridge, MA: MIT Press 2020.

surprising that repair has emerged only recently as a major field of research for historians of technology and scholars of science and technology studies. One reason why maintenance and repair have been overlooked can be found in the traditional innovation-centric research agenda of the history of technology. This is why the current "maintainers network" argues for an emphasis on maintenance instead of the traditional focus on invention and innovation in the field.[5]

In the research literature on repair that has become available, it is possible to identify four clusters, each of which maps out a distinct way of looking at repair: first, "broken world thinking"; second, repair as invisible work; third, repair as *bricolage*; and last, repair as an innovative act.

The first cluster presents a call to consider any technological act from the starting point of brokenness. Steve Jackson recently argued for "broken world thinking": historians of technology should take "erosion, breakdown, and decay, rather than novelty, growth, and progress, as ... starting points" for their research and narratives.[6] Obviously, repair then takes centre stage as it serves to remediate decay and breakdown.

In the second cluster, repair is described as a "hidden field" or as "invisible work". In this view, repair is conducted behind the scenes – in an unseen backroom, for example, in the case of consumer goods or, in the case of infrastructure, at night, when it causes the minimum possible disruption. In addition, repairs are often carried out in informal markets, a phenomenon that has best been described in the context of the Global South. It is also becoming clear that the low social status frequently assigned to repairers contributes to their apparent invisibility.[7]

The third cluster examines the particular characteristics of repair know-how. This know-how is conceptualised as experiential, situational and embodied knowledge. Improvisation and *bricolage* are emphasised.[8] This literature also

---

5   See themaintainers.org (accessed 04.11.2019), see also Russell, Andrew L./Vinsel, Lee: "After Innovation, Turn to Maintenance", in: Technology and Culture 59, 1 (2018), p. 1–25.

6   Jackson, Steven J.: "Rethinking Repair", in: Gillespie, Tarleton/Boczkowski, Pablo J./ Foot, Kirsten A. (eds.): Media Technologies: Essays on Communication, Materiality, and Society, Cambridge, MA: MIT Press, 2014, p. 221–239, here p. 221.

7   See e.g. Henke, Christopher: "The Mechanics of Workplace Order: Toward a Sociology of Repair", in: Berkeley Journal of Sociology 44 (1999/2000), p. 55–81.

8   See e.g. Orr, Julian: Talking about Machines. An Ethnography of a Modern Job, Ithaca, NY/London: Cornell University Press 1996; Strebel, Ignaz/Bovet, Alain/Sormani, Philippe (eds.): Repair Work Ethnographies, Singapore: Palgrave Macmillan 2019.

provides some insight into the difficulties involved in any attempt to standardise and automate maintenance and repair work.

The fourth cluster stresses the innovative nature of repair, despite the fact that it is essentially about conserving existing things or infrastructures. Repairers often alter the original structure of things, textiles or other objects; such repairs are described in this literature as incremental innovations.[9] Another aspect is the blending of old and new, of "Western" and "non-Western" technology as a result of repairs – what David Edgerton, for example, has termed "creole technologies".[10]

Adding to this literature, we argue in this volume that repair should be discussed from a temporal perspective – one which reaches beyond the timescale of the repair process itself.[11] This includes the historicity of repair, i. e. that repair practices and cultures have changed over time and should be investigated in their respective historical contexts. But it reaches beyond historicity and refers to the manifold temporalities included in processes, infrastructures and acts of repair. Maintenance and repair react to the wear and tear that happens over time and they represent interventions with the temporal aim of prolonging the time an object can stay in use. Moreover, when studying repair we also need to raise the question of what comes after repair (or non-repair): removing or hoarding for future reuse or care? Reuse, e.g. through second-hand resale or dismantling into reusable parts? Sorting for final disposal? And what happens thereafter?

Our claims are thus twofold: firstly, repair practices should be viewed from a historical perspective. Our selection of case studies in this volume focuses on seminal 20th-century technologies, from infrastructure to production plants to the motor car. It is often assumed that practices of repair and reuse have gradually declined along with the rise of 20th-century mass production, mass consumption and throwaway societies.[12] History shows, however, that repair has always gone hand in hand with any human-object interaction, from mediaeval bridges to

---

9   See e.g. Graham/Thrift, "Out of Order", p. 5.

10   Edgerton, David: "Creole Technologies and Global Histories: Rethinking how Things Travel in Space and Time", in: Journal of History of Science and Technology 1, 1 (2007), p. 75–112.

11   On different timescales of repair (e.g. repair as routine activity, the before and after of repair, preventive maintenance, etc.), see Henke/Sims: Repairing Infrastructures, p. 25–26.

12   For a more detailed account, see Krebs, Stefan/Weber, Heike: "Rethinking the History of Repair: Repair Cultures and the 'Lifespan' of Things" (this volume).

today's mobile phones.[13] Technological infrastructures and technology are all about serviceability and usability – which, of course, implies maintenance and, if there are faults, repair. This is why we argue that the declensionist narrative does not reflect the historical development of repair. A closer historical look at 20th-century infrastructures and consumer cultures demonstrates that maintenance and repair have not become obsolete in modern consumer societies; rather, their modes, their appearance and the sites and actors of repair have changed substantially.[14] However, we still know surprisingly little about these historical changes and even less about what happens after repair. In his plea for a history of "technology-in-use", David Edgerton summarised: "Unfortunately we are not in a position to give an overview of the main trends in the history of maintenance and repair. Has maintenance as a proportion of output gone up or down? Where there has been a trade-off between initial cost and maintenance, what have producers and consumers gone for?".[15] This finding has changed little in the past ten years. More recent historical studies, including this volume, suggest that repair practices have followed different trajectories and trade cycles.[16] Krebs and Hoppenheit have for example shown that employment in the repair sector continued to increase in the 1970s and 1980s.[17] The overall importance of repair might have declined for some (consumer) technologies, but emerging technologies have also led to the establishment of new fields of maintenance and repair. The widespread adoption of cars, radios and washing machines, for instance, was based on customer services and repair facilities and on second-hand markets. Furthermore, repair knowledge, tools and motives have also changed over time.

Second, we argue that practices of maintenance and repair are not only linked to the innovation, use and consumption of technology but that they are part and parcel of technology's different temporalities. When technical artefacts become old and worn out, their users or owners have to decide whether it is necessary, worthwhile or possible to maintain and repair them and thus extend their

---

13 On smartphone repair stores in Switzerland, see Nova, Nicolas/Bloch, Anaïs: Dr. Smartphone: An Ethnography of Mobile Phone Repair Shops, Lausanne: IDPURE 2020.

14 See Krebs, Stefan/Hoppenheit, Thomas: "Questioning the Decline of Repair in the Late 20th Century: the Case of Luxembourg, 1945-1990", in: Hilaire-Pérez, Liliane et al. (eds.): Technical Cultures of Repair from Prehistory to the Present Day, Turnhout: Brepols Publishers 2021 (forthcoming).

15 Edgerton, David: The Shock of the Old, London: Profile Books 2006, here p. 81.

16 For a more detailed account, see Krebs/Weber, "Rethinking the History of Repair".

17 Krebs/Hoppenheit: "Questioning the Decline of Repair".

present use; to reuse, hoard or dismantle them for different purposes; or finally to get rid of them. So we need to tackle the question of what becomes of "old" technologies: whether they are repaired or not is closely related to questions of reuse and removal, dismantling and disposal – the complex decision between "mending and ending" (see also Weber, this volume) depends on many factors including the availability of second-hand markets, repair infrastructures and dismantling or disposal facilities.

Accordingly, in this volume, we intend to go a step beyond "broken-world thinking". For us repair is one of the many aspects of the temporalities of technology, and in particular of its intractable persistence, which rarely ends with the end-of-use of a technological artefact. The questions of how long and in what shape technology remains in use, how and why it is taken out of use and what happens afterwards are related to the contexts and conditions of maintenance, repair, reuse and disposal infrastructures, to their availability or absence, and to the related economies of waste, recycling and reuse.

In the first part of this introduction, we want to elaborate on what we mean by temporalities of technology, including the "persistence of technology" which gave the book its title. The second part provides an overview of the contributions in the volume, all of which stress questions of repair, reuse and disposal in situations of technology-in-use, technology-in-the-making or technology-in-the-unmaking. The chapters focus on technologies which have shaped the 20th century such as the power grid, ocean-going vessels, telephones and cars. Geographically, we cover various Eastern and Western European countries, North America, China, India and the former Soviet Union.

TECHNOLOGY'S PERSISTENCE: A TEMPORAL PERSPECTIVE ON TECHNOLOGY AND ITS REPAIR, REUSE AND DISPOSAL

In the field of history, considerable thought, and some rethinking, is currently being devoted to the subject of time and temporality.[18] By bringing such thoughts into the history of technology, we argue that technology harbours manifold temporal dimensions, including the fact that it is relatively persistent.

---

18 See e.g. the "Viewpoints" section in Past & Present 243, 1 (2019); Champion, Matthew: "The History of Temporalities: An Introduction", in: Past & Present 243, 1 (2019), p. 247–257; Tanaka, Stefan: "History without Chronology", in: Public Culture 28, 1 (2015), p. 161–186.

From such a temporal perspective, repair becomes an intervention that is intended to prolong the time that a certain technology can stay in use; dismantling and disposal, by contrast, are interventions that bring an end to the use phase. In most cases, these interventions disaggregate and transform infrastructures, buildings or things into reusable parts, recyclable materials and "stuff" to be discarded, and they often come with an "afterlife", as demonstrated by the issues of waste legacies or industrial ruins.

As contemporaries of the COVID-19 crisis, we have all been eyewitnesses to diverse temporalities of technology: some infrastructures were brought to a standstill, others were accelerated. Hospitals were erected in just a few weeks, while the grounding of around 20,000 planes entailed complex caretaking activities to stockpile them for future use. Intercontinental container shipping schedules were disrupted, resulting in blank sailings (cancellations) and prompting the growing practice of "schedule sliding" (adding buffer time to sailing schedules to allow for delays). Reflections on technology and its development in respect to temporalities – which go far beyond temporal issues such as the chronologies and innovation timelines of technologies or the technical measurement of time – are beginning to emerge in science and technology studies and the history of technology. Heike Weber, for instance, has appropriated Reinhart Koselleck's metaphor of "Zeitschichten" (sediments or layers of time) to map out the timescapes of technology, and Jens Ivo Engels has interpreted technical infrastructures as products as well as producers of time.[19] From a media studies perspective, Gabriele Schabacher has argued that any infrastructure is formed by layers of different ages and follows temporally different patterns of (non-)care such as repair, abandonment or repurposing, while Gabriele Balbi and Roberto Leggero underline that focusing on maintenance can help us to understand communication infrastructures in their *longue durée* existence.[20] Similarly, anthropologists have conceptualised infrastructure as a process over time, from

---

19 Weber, Heike: "Zeitschichten des Technischen: Zum Momentum, 'Alter(n)' und Verschwinden von Technik", in: Heßler, Martina/Weber, Heike (eds.): Provokationen der Technikgeschichte. Zum Reflexionsdruck historischer Forschung, Paderborn: Schöningh 2019, p. 107–150; Engels, Jens Ivo: "Infrastrukturen als Produkte und Produzenten von Zeit", in: NTM. Zeitschrift für Geschichte der Wissenschaften, Technik und Medizin 28, 1 (2020), p. 69–90.

20 Balbi, Gabriele/Leggero, Roberto: "Communication *is* maintenance: turning the agenda of media and communication studies upside down", in: H-ermes. Journal of Communication 17 (2020), p.7–26.

design to construction then maintenance or abandonment, breakdown, demolition or ruin.[21]

In this collection, we want to highlight three temporal dimensions:[22] first, the polychronic structure of our mechanised world, i. e. that any given society has used or uses technologies from diverse past times and from the present; second, each technology goes along with certain temporalities ascribed to or inscribed in it, e.g. expectations on how long it should be in operation; third, these temporalities include the so-called "afterlife" of technology as one of the most problematic examples of its potential persistence. These are temporal dimensions for which historians of technology have yet to develop a keen awareness – an awareness that extends beyond industrial archaeology or the conservation and restoration of historical objects which up to now have constituted the main fields for history's reasoning on "aged" technology. They ultimately concern hitherto overlooked questions, namely how societies value and treat a technology in respect to time and how societies not only "make" and "use" technology but also "unmake" it.

By referring to the "polychronic" nature of technology, we want to underline that in no society has the latest technology been used to the exclusion of older technologies; on the contrary, the technology of any given historical period has always been a patchwork of old and new. Most technologies persist long after the emergence of technically superior alternatives or are put to other uses elsewhere. David Edgerton in particular has underlined this point recently with his technology-in-use perspective: the objects, infrastructures and practices in use originate in different historical times, yet they coexist simultaneously and in parallel in our present. Svante Lindqvist made a similar point in the mid-1990s. The world of technology, he said, was almost entirely driven by "old-age" technologies that had already reached maturity or were in decline. "For any given technology and at any time we will find that the prevailing technological volume is a mixture of several and at least the following three components: an older technology in decline (A), a second at its peak (B), and a third one emerging (C)."[23] The

---

21 Schabacher, Gabriele: "Time and Technology: The Temporalities of Care", in: Volmar, Axel/Stine, Kyle (eds.): Hardwired Temporalities. Media, Infrastructures, and the Patterning of Time, Amsterdam: Amsterdam University Press (forthcoming); Anand, Nikhil/Gupta, Akhil/Appel, Hannah (eds.): The Promise of Infrastructure, Durham/London: Duke University Press 2018.

22 The following paragraphs draw on Weber, "Zeitschichten des Technischen".

23 Lindqvist, Svante: "Changes in the Technological Landscape. The Temporal Dimension in the Growth and Decline of Large Technological Systems", in: Granstrand, Ove

illustrative examples he cites include the last charcoal-fired blast furnace in Sweden, which was not closed down until 1966, and the use of horses in Germany, which, because of their enduring importance in agriculture, did not peak until the 1920s, in other words when motor vehicles were already on the rise. In his book *The Shock of the Old*, David Edgerton identifies a multitude of other examples, including "shocking" cases such as asbestos, which is still a common construction material in many parts of the world (see also Dhawan, this volume). This polychronicity is in evidence even in the context of one single technological system in a given region. Today's automobility, for instance, comprises a constantly evolving diverse vehicle pool and myriads of infrastructural elements that require constant maintenance and renewal. Likewise, the "digital revolution" of our day would have been impossible without the copper-core cables of the telephone era; the late 20th-century German telecommunication system, for instance, identified copper cables rather than optical cables as the basis for future digitisation.[24] In view of this polychronicity of the technical world, it would be wise to shed the dualistic conception of "old" and "new" technologies altogether,[25] because this terminology suggests a linear sequence or even replacement – often associated with linear progress – , whereas an additive overlap and a polychronic hybridity actually prevail. The manifestations of this technological polychronicity vary between regions and historical eras and they essentially depend on the respective cultures of maintenance, repair, reuse and disposal.

The second aspect concerns temporal dimensions ascribed to or inscribed in technology itself, e.g. innovation cycles or questions of durability, degeneration and obsolescence or the persistence of technology. A current pertinent example is "Moore's law", which has dictated the short innovation cycles of digital equipment in recent decades.[26] We often apply anthropomorphic metaphors such

---

(ed.): Economics of Technology, Amsterdam: Elsevier 1994, p. 271–288, here p. 276 and 284.

24 Henrich-Franke, Christian: "'Alter Draht' – 'neue Kommunikation': Die Umnutzung des doppeldrahtigen Kupferkabels in der Entwicklung der digitalen Telekommunikation", in: Habscheid, Stephan et al. (eds.): Umnutzung: Alte Sachen, neue Zwecke, Göttingen: V&R unipress 2014, p. 97–112.

25 For a more detailed treatment, see Weber, "Zeitschichten des Technischen"; see also: Tanaka, History without Chronology.

26 While Ceruzzi framed Moore's Law as technological determinism, Mody hinted at the social construction of this "law", e.g. through mass sales of laptops and cell phones. See Ceruzzi, Paul E.: "Moore's Law and Technological Determinism: Reflections on the History of Technology", in: Technology and Culture, 46, 3 (2005), p. 584–593;

as "age", "lifespan", "technological generations" or "death" to refer to some of these temporal dimensions of technology, as if the "biography of things" resembled a human biography. While Arjun Appadurai's idea of an "anthropology of things" was once helpful in understanding changing social and cultural meanings of objects,[27] in the light of recent challenges such as fast fashion, increasing quantities of waste and the toxicity of e-waste, it seems inadequate to parallelise the temporalities of the material word with the traditional and centuries-old idea of human life stages. Yet in the absence of more appropriate terms, our book also sometimes applies terms such as "old" technology or the "afterlife" of technology.

It is therefore all the more important to emphasise that technological objects rarely follow the plain "bio-narrative" of an arrow-like path from cradle to coffin. Large-scale infrastructure systems such as telephone networks or electricity grids (see the contributions by Tan, Lean and Hadlaw in this volume), for instance, are highly polychronic entities, which are never switched off unless a fault or blackout so dictates. Servicing, overhaul, refurbishment, repair and constant updating or replacement of outdated parts are indispensable for these systems to deliver the desired continuous operability, and these processes serve to counteract the infrastructure's inevitable degeneration and wear and tear. The case of airports and planes during the coronavirus crisis has demonstrated that infrastructures cannot simply be put on hold; grounding aircraft requires time-sensitive activities such as preparing engines and tanks so that they can be put into storage. By contrast, other technologies or infrastructures are simply left to decay and turn to ruin, or are dismantled or demolished. Some technological artefacts remain operational through diverse cascades of reuse; others are salvaged for various reasons or conserved as cultural treasures in museums, by private collectors or hobbyists. Certain old car models, for instance, have become objects of such intentional preservation by hobbyists that they serve as a kind of "time capsule" (see Lucsko in this volume).

When it comes to the "lifespans" of technical artefacts, material wear and tear and especially – at least in mass consumer societies – cultural obsolescence and society's expectations on technical progress and newness define whether, when and why artefacts are to be considered as "aged" and "obsolete". The example of houses illustrates the strong influence that regional cultures have on

Mody, Cyrus: The Long Arm of Moore's Law. Microelectronics and American Science, Cambridge, MA: MIT Press 2017.

27 Appadurai, Arjun (ed.): The Social Life of Things. Commodities in Cultural Perspective, Cambridge et al.: Cornell University Press 1986.

construction: an average house in Japan, for instance, stands for 30 years before it is demolished; in the United States, this average lifespan amounts to 55 years and, in Britain, no less than 77 years. Needless to say, with these diverse lifespans come differing practices, intensities and costs of maintenance and repair,[28] as demonstrated by the case of Samarkand houses and their cross-generational persistence based on tacit construction knowledge and continuous repair and restoration (see van der Straeten/Petrova in this volume).

The idea of quantifiable "service lives" of technology emerged along with the development of mass production and mass consumption, and it is a notion that forms the backbone of any planning and engineering of investment and consumer goods.[29] Twentieth-century bridges, for instance, were designed to last for several decades, whereas the product lifespan of cars, based on average use frequency and driving habits, was conceptualised within a range of 10 to 12 years. Currently, a mobile phone is considered outdated after less than two years, and Snapchat is programmed to obliterate messages after 24 hours. As these examples show, by the late 20th century time and technology had intersected in novel forms that were not designed to coordinate the rhythms of work, workers and machines as in the 19th century, but to shape institutional, organisational and social time constructs for innovation and allow "adequate" time for substitution and the decline of "mature" technology.[30]

Many technological objects that are discarded in wealthy societies as "obsolete", however, find their way, after repair or dismantling and resale, into second-hand markets. In contrast to pre-modern markets, these modern second-hand markets and their scope and meaning for mass consumption are as yet underexplored,[31] but it is clear that they have played a key role, for example in the dissemination of consumer technologies to less affluent consumer classes. In the course of the 20th century, trade routes lengthened from local to global, with the

---

28 Cairns, Stephen/Jacobs, Jane M.: Buildings must die: a perverse view of architecture, Cambridge, MA et al.: MIT Press 2014, p. 127.

29 See Weber, Heike: "Made to Break? – Lebensdauer, Reparierbarkeit und Obsoleszenz in der Geschichte des Massenkonsums von Technik", in: Krebs/Schabacher/Weber (eds.), Kulturen des Reparierens: Dinge – Wissen – Praktiken, Bielefeld: transcript 2018, p. 49–83; Slade, Giles: Made to Break: Technology and Obsolescence in America, Cambridge, MA: Harvard University Press 2006.

30 Helga Nowotny has spoken of a "chrono-technology" for these processes, see Nowotny, Helga: Eigenzeit. Entstehung und Strukturierung eines Zeitgefühls, Frankfurt a. M.: Suhrkamp 1989, p. 64–66 and 73.

31 Fontaine, Laurence (ed.): Alternative Exchanges. Second-Hand Circulations from the Sixteenth Century to the Present, Oxford/New York: Berghahn Books 2008.

result that used technology is now primarily exported from Western places of first use to poorer regions of the Global South, and often through informal channels and markets.

While historical research on the polychronic structure of technology, on user cascades, second-hand markets or obsolescence is relatively rare, virtually no historical studies have thus far considered the "unmaking" of technology. Researchers in the fields of social science and the history of technology have filled libraries with concepts on innovation and the "making" of technology, and they have developed notions such as appropriation, domestication, normalisation or creolisation to describe the use phase. By contrast, there is a conceptual absence when it comes to the removal, dismantling, decline and decay of technology. Only some initial articles have explored what the vast field of "unmaking" of technology could mean for the history of technology.[32] Technology does not simply disappear; every removal from the place of service or use requires an active intervention, once the decision for "divestment"[33] has been taken.

Studies within the field of consumption history on throwaway practices might generate valuable input. For example, in her seminal book *Waste and Want*, Susan Strasser explained how in American households the once common "stewardship of objects" was gradually replaced by a throwaway culture, once mass consumption and municipal disposal infrastructures were taken for granted.[34] Technologies of removal, however, include more than waste disposal services. Indeed, alongside the repair sector, there exist dismantling and scrapping businesses – a vastly underexplored and often informal, even illegal field of economic activity which has been analysed for the metal scrap business and car recycling.[35] More attention has been paid to cities and their inherently polychronic

---

32 Rare examples are Weber, Heike: "'Entschaffen': Reste und das Ausrangieren, Zerlegen und Beseitigen des Gemachten (Einleitung)", in: Technikgeschichte 81, 1 (2014), p. 1–32; Salehabadi, Djahane: "The Scramble for Digital Waste in Berlin", in: Trischler, Helmuth/Oldenziel, Ruth (eds.): Cycling and Recycling. Histories of Sustainable Practices, Oxford/New York: Berghahn Books 2016, p. 202–212; Zimring, Carl A.: "The Complex Environmental Legacy of the Automobile Shredder", in: Technology and Culture 52, 3 (2011), p. 523–547.

33 Gregson, Nicky/Metcalfe, Alan/Crewe, Louise: "Moving things along: the conduits and practices of household divestment", in: Transactions of the Institute of British Geographers, 32, 2 (2007), p. 187–200.

34 Strasser, Susan: Waste and Want. A Social History of Trash, New York: Metropolitan Books 1999.

35 Zimring, The Complex Environmental Legacy; id.: Cash for your Trash. Scrap Recycling in America, New Brunswick/London: Rutgers University Press 2005; Denton,

architecture: demolition, obduracy and rebuilding have shaped modern cities and are part of urban planning.[36]

The challenge of "unmaking" technology is encapsulated in the numerous 21st-century photographs of plastic or e-waste piles and industrial ruins of our age, such as Edward Burtynsky's "technofossils" series.[37] These images also remind us of the third point, the "afterlife" of technology. Historians have yet to begin to examine the diverse legacies of "unmaking" technology – after-effects that extend into an unknown future in which the technology itself might no longer exist in its present form. While the long-term effects of technology are nothing new, by the late 20th century they had assumed unprecedented geographical and temporal dimensions – dimensions currently explored in the ongoing Anthropocene debate.[38] The 21st-century global world is increasingly lacking so-called sinks in which to dispose of all the extracted resources and manufactured products, while toxic waste legacies – such as microplastics in the sea or in our blood or the growing levels of carbon dioxide in the earth's atmosphere – have rebutted the idea of an "ultimate sink" with which we could forever unmake the made.[39] Besides, many engineering activities are intended to "repair" technological paths taken by former generations: examples range from the insulation of buildings to reduce heating energy to the latest order on phosphate recovery from sewage sludge – measures that are intended to lighten the ecological footprint of the given infrastructure; likewise, electric cars are meant to remedy the damage caused by emissions from combustion engines (see Marhold in this volume). Moreover, remediation and so called "after-care" have become genuine fields of interaction between humans and technology with the aim of "repairing" the after-effects of certain technologies. In coal areas, for instance, water regula-

---

Chad/Weber, Heike: "Rethinking Waste within Business History: A Transnational Perspective on Waste Recycling in World War II.", in: Business History, DOI: 10.10 80/00076791.2021.1919092.

36 Hommels, Anique: Unbuilding Cities. Obduracy in Urban Sociotechnical Change, Cambridge, MA/London: MIT Press, 2008; Ryan, Brent D.: Design after Decline. How America Rebuilds Shrinking Cities, Philadelphia: University of Pennsylvania Press 2012.

37 Burtynsky, Edward/Baichwal, Jennifer/de Pencier, Nicholas: Anthropocene, Göttingen: Steidl 2018.

38 See e.g. Bonneuil, Christophe/Fressoz, Jean-Baptiste: L'Evénement Anthropocène. La Terre, l'histoire et nous, Paris: Édition du Seuil 2013.

39 Tarr, Joel A.: The Search for the Ultimate Sink. Urban Pollution in Historical Perspective, Akron: University of Akron Press 1996.

tion and evacuation work are in place and remain so even after the mining activity itself has been abandoned.

With this volume, we want to demonstrate that a temporal perspective on technology and its persistence has an important role to play in the history of technology. In our view, the "Shock of the Old" (D. Edgerton) is not only about the long and diverse technology-in-use phase; it is also about the fact that technologies remain efficacious even beyond that phase. The notion includes situations of repair and reuse as much as technology's abandonment, decay or removal and diverse forms of its "afterlife". It is only by taking this persistence of technology seriously that we can appreciate how closely the practices, structures and economies of repair, reuse and removal are interwoven and understand how they have changed over time.

FROM MAINTENANCE AND REPAIR TO REUSE AND DISPOSAL: THE SECTIONS OF THE BOOK

Repair has always been a dominant field of interaction between humans and their technologies. Production and infrastructure facilities are in constant need of maintenance to keep them running. And even the spread of new consumer technologies such as automobiles, television sets and household appliances has greatly depended on maintenance and repair services as well as second-hand markets and refurbishment shops. In our introductory essay "Rethinking the History of Repair" (Krebs/Weber, this volume) we question the common narrative of a linear decline of repair during the 20th century. Instead we argue that the long history of repairing things saw multiple ups and downs, with changing cultures of repair and DIY repair and a varying set of actors involved. Moreover, changing disposal infrastructures and changing practices of reuse and disposal have shaped the forms and intensities of repair.

As already mentioned, infrastructure such as roads and electricity grids requires constant maintenance and repair. The first section of the book, "Maintaining Infrastructures", brings together three chapters that feature different times and geographical regions. In the first chapter, Ying Jia Tan investigates the repair of China's power grid between the Anti-Japanese Resistance and the early years of the People's Republic. During this time of what he calls "perpetual warfare" a significant shift in repair culture occurred. The system builders of the Chinese electrical network were initially preoccupied with the replacement of worn-out and defective machinery. Cut off from foreign supplies, Chinese engineers had to turn to repair to make war-damaged turbines run again. During the early years of the People's Republic two different repair cultures competed with each other: a more systematic top-down engineering approach, and more bottom-up repair practices of ordinary workers and technicians. The latter fitted well with the Party's ideology of mass mobilisation, highlighting the political significance of maintenance and repair.

The next chapter deals with another political decision that significantly changed repair cultures. In "Changing Perceptions of Repair and Maintenance", Thomas Lean describes how repair practices and perceptions of them changed after the privatisation of the British electricity supply industries. During the state monopoly period, maintenance and repair were given high priority. Drawing on oral history interviews Lean shows that engineers identified with high maintenance standards because they were delivering a public service aimed at "keeping the lights on" in the country. After privatisation a more flexible and supposedly also more efficient maintenance regime was introduced to ensure the financial profitability of the now private electricity suppliers. Although the lights stayed on, engineers struggled with the new repair practices, which were inconsistent with their identity of delivering public service.

Under the title "Business as Usual", Jan Hadlaw investigates in the third chapter maintenance and repair at Bell Telephone Company of Canada. Between the 1880s and the 1930s, the North American telephone market was divided between several companies. For these monopolistic providers the "telephone plant" encompassed everything from local loops, trunks, switches and cables to telephones in the homes and offices of their customers. Because they owned the telephone sets and private branch exchanges, maintenance and repair of the equipment was an integral part of the companies' operations. In the mid-1920s, when the first American manufacturers turned towards planned obsolescence, Bell Canada decided to expand and rationalise its repair activities. However, for Bell it was an economic rather than an environmental concern to keep its telephones in service for as long as possible.

The chapters by Lean and Hadlaw highlight how business cultures of state utilities and private companies had a decisive influence on maintenance and repair practices. They also reveal that repair regimes shaped the perceptions and identities of technicians and engineers, and that changes in repair standards challenged the self-perception of the workforce. Furthermore, Tan and Lean show the interconnectedness of political ideologies and repair: while Maoist communism enforced bottom-up repair practices, the neo-liberal privatisation of the British electricity network favoured a new short-term maintenance regime. However, the "Maintaining Infrastructures" section also reminds us that the history of technology is in need of more systematic and comparative studies on how the maintenance, replacement and disposal of technologies in large technical systems developed over time.

The second section focuses on "Users and Repair" – house owners, car owners and professional car and bus drivers and their activities of repairing, reworking or improving technology. In "Building, Maintaining and Improving One's Own House in Soviet Samarkand", Jonas van der Straeten and Mariya Petrova explore the case of self-making and analyse building, maintenance and repair strategies for adobe courtyard houses in Samarkand. During the Soviet period and its manifold modernisation efforts, this house type, associated with private house ownership, was dominant, along with the respective traditional practices and knowledge of constructing and repairing it; prefabricated concrete apartment blocks made their appearance only at the urban margins. While the construction of adobe courtyard houses required high levels of labour and the finished houses were in need of constant care and rebuilding – not least each year after winter –, they suited local needs, mentalities and customs. In the regional tradition of collective self-help and labour mobilisation (*hashar*), male relatives or neighbours participated in the (re)building processes, while females took over caring activities such as preparing meals for everyone. By preserving their traditional material environment through constant building and repairing, the residents of Samarkand also maintained their pre-Soviet cultural identity in Soviet times.

In his chapter "Maintaining the Mobility of Motor Cars: the Case of (West) Germany, 1918-1980", Stefan Krebs investigates maintenance and repair as a central part of automobility. Mobility has always been (and still is) at the heart of car consumption. A motor car would lose its use value as a consumer item, at least temporarily, in the event of a breakdown, and the exchange value of a broken car would be lower if it were sold. So maintenance and repair to prevent or remedy malfunctions were necessary and recurrent moments in the consumption of an automobile. However, car repair also became a leisure activity in the interwar years and especially in the post-war period, when members of the working classes started to own automobiles. Self-repair was cheaper than taking a car to a professional garage. Furthermore, self-repair also served as means to shape and foster male identities as skilled and knowledgeable amateur mechanics.

As Karsten Marhold demonstrates in his article "Of Buses, Batteries and Breakdowns: The Quest to Build a Reliable Electric Vehicle in the 1970s", engineers from a German and a French electrical utility company saw battery maintenance as a major challenge in constructing an electrical car for daily use. For them, reliability ranked higher than performance, and batteries were the key issue here since charging and changing batteries required regular inspections and knowledgeable care. However, drivers tended to charge and discharge the batteries of their electric car in sub-optimal ways, a fact which further complicated the battery maintenance issue. While questions of maintenance thus shaped innovation processes, engineers did not take into account the battery's afterlife and disposal, even if contemporary batteries had relatively short life cycles. Dismantling and disposal were and still are barely taken into account in innovation and production.

The third section of the book, "Reuse and Conservation", looks at two examples of the persistence of apparently obsolete technology. In "A Bargain or a 'Mousetrap'? A reused Penicillin Plant and the Yugoslavians' Quest for a Healthier Life in the Early Post-war Era", Sławomir Łotysz investigates the transfer of a Canadian Merck penicillin plant to Yugoslavia – a plant that was actually worn out and had an obsolete design. Łotysz scrutinises the different arguments of the Yugoslavian officials and engineers who insisted on acquiring a second-hand plant instead of accepting an offer from UNRRA for new equipment. This historical case also highlights the long "lifespan" and persistence of technology in basic and heavy industries.

David Lucsko studies a completely different case: that of old cars. In the chapter "'Proof of Life': Restoration and Old-Car Patina", he traces the history of a new trend in old car restoration. For many years old car enthusiasts tried to refurbish their cars to factory-new conditions, but since the early 2000s some of them have started to proudly display the faded paint, patches of rust and worn and stained upholstery of their restored cars. In this example the persistence of old technology is, of course, driven by very different motives as these cars became fashionable and precious collectibles long after they had become obsolete and were put out of use. The paradox identified by Lucsko is that the "patina" cars are carefully repaired "time capsules" that are perceived by their owners as being more original than their "factory-new"-restored counterparts, despite the fact that they were obviously not worn out when they were new.

Both chapters highlight that the persistence of technologies and their potential cascades of use depend not only on economic and technical factors but also on cultural and ideological circumstances. The history of obsolete technologies that are still used and valued by some actors can help improve our understanding of the different temporalities of technological objects.

The final section, "Obsolescence and Disposal", sheds in its two chapters light on the close interrelatedness of production, repair, reuse, recycling and removal. It tackles the question of what comes after repair and why the option of repair is sometimes rejected when decisions for final removal are taken. Moreover, the section underlines that the temporalities of technology do not end with removal and that technology's "unmaking" might have an "afterlife".

In her article "Mending or Ending? Consumer Durables, Obsolescence and Practices of Reuse, Repair and Disposal in West Germany (1960s–1980s)", Heike Weber tackles changing practices of "mending" and "ending" in West Germany through the lens of contemporary bulk waste collections, repair services and popular repair booklets, and the planned obsolescence debate that erupted in the early 1970s. By the 1960s, the FRG had turned into a mass consumer society. In the ensuing years, the American model of a "throwaway society" was widely criticised, and in the 1970s an environmental awareness took hold, at a time when the public discourse also included major criticism of the "planned obsolescence" idea. Nevertheless, it was during this critical period that West German consumers considerably changed their practices of care and divestment with respect to consumer durables. Repair and reuse did not disappear, but the majority of consumer durables were eventually sorted and discarded via bulk waste collections – thereby redefining the "durability" implicit in the term "consumer durables" as merely a modest number of years.

Diverse cascades of use, even forms of reuse at the stage of dismantling and disposal, are highlighted in the contribution by Ayushi Dhawan, who also reflects on issues of interregional transfer and the afterlife of technology. In her article "The Persistence of *SS France*: Her Unmaking at the Alang Shipbreaking Yard in India", she takes a close look at the widely discussed topic of ship dismantling which for nearly all the world's ships happens on a few stretches of the Indian and Bangladeshi coastlines. Ships often have a very long lifespan and Ayushi Dhawan follows the example of the *SS France* through her many stages of life, starting with her initial use as a famous French ocean liner and her reuse as a Caribbean cruise ship (as the *SS Norway*) and ending with her scrapping as the *SS Blue Lady* in the Alang Shipbreaking Yard. Her story contains many sad ironies. The ship was – illegally – sent to be scrapped after a boiler accident, a consequence of both material fatigue and poor, even careless maintenance and inspection. While her legal decontamination and disposal in accordance with the European Waste Shipment Regulations would have incurred high costs, Indian court authorities decided that her unmaking at the Alang Yard would provide both jobs and reusable asbestos for local building purposes. So although the life of the ship thus came to an end, parts of the *SS France* still persisted – whether in the form of reused steel, cutlery, clocks or fire extinguishers sold at the local second-hand markets or waste dumped at the local landfill. Dhawan sums up that the arduous unmaking represented a process that was "toxic and life-giving at the same time".

Rethinking the History of Repair:

Repair Cultures and the "Lifespan" of Things

Stefan Krebs and Heike Weber

The act of repair is inherent to everyday life.[1] Repair processes are inextricably linked not only to the things we repair, but also to infrastructures and organisational processes. Repair practices therefore play a vitally important role in reciprocal interactions between humans and technology – a role associated with highly specific knowledge about things, equipment, processes and interventions. Steven Jackson describes his article "Rethinking Repair" as an exercise in "broken world thinking".[2] In so doing, he employs a methodological principle which Geoffrey Bowker has coined as "infrastructural inversion",[3] i. e. "learning to look closely at technologies and arrangements that, by design and by habit, tend to fade into the woodwork".[4] When Jackson examines the act of repair, he therefore posits that the normal state is for a system to be fundamentally broken rather than for it to be functioning properly. Only from this perspective does it become clear that certain systems (in particular major infrastructure systems such as gas,

1 This is a translated and abridged version of Krebs, Stefan/Schabacher, Gabriele/Weber, Heike: "Kulturen des Reparierens und die Lebensdauer der Dinge", in: id. (eds.): Kulturen des Reparierens: Dinge – Wissen – Praktiken, Bielefeld: transcript 2018, p. 9–46.

2 Jackson, Steven J.: "Rethinking Repair", in: Gillespie, Tarleton/Boczkowski, Pablo J./Foot, Kirsten A. (eds.): Media Technologies. Essays on Communication, Materiality, and Society, Cambridge, MA/London: MIT Press 2014, p. 221–239, here p. 221.

3 Bowker, Geoffrey: "Information Mythology. The World of/as Information", in: Bud-Frierman, Lisa (ed.): Information Acumen. The Understanding and Use of Knowledge in Modern Business, London/New York: Routledge 1994, p. 231–247.

4 Bowker, Geoffrey/Star, Susan Leigh: Sorting Things Out: Classification and its Consequences, Cambridge, MA/London: MIT Press 1999, p. 34.

telephone, underground railways, etc.) can be kept up and running solely through never-ending practices of repair and maintenance, since all defects, up to and including system failure, are inherent to them.[5]

Repairs are often unscheduled and arise out of a need to eliminate faults and to make things which have broken down useful again. These criteria distinguish repairs from maintenance, which is a precautionary activity that is generally scheduled in advance. Nevertheless, maintenance and repair tasks do have some points in common – they both postpone the day when a thing wears out to the point at which it becomes unusable, and they both therefore influence the question of wear and tear and the "lifespan" of things, or in other words how long a thing or good is used for and when it should be removed from circulation.

The purpose of this chapter is to historicise the concept of "repairing things" with a view to broadening and redefining the emphasis of current debates on repair as a "new social movement" and the emergence of a "repair society". Such discourses often allude to the technical empowerment of citizens, "convivialism" and sustainability. For example, Andrea Baier et al. regard certain elements of the current repair and do-it-yourself movement as a "post-capitalist practice" characterised by three different features: (1) an ethical interest in opposing capitalism through subsistence, participation, benevolence and post-growth; (2) the sharing of things ("Do It Together" (DIT) as part of Do It Yourself (DIY)); and (3) general access to repair knowledge, promoted by the Internet as a digital commons which makes knowledge that was previously the purview of a few available to everyone.[6] Indeed, representatives of the repair movement explicitly view themselves as allies in the fight "against short product lifespans".[7] What is more, they interpret repairs themselves as an expression of growing technical literacy. Wolfgang Heckl also argues in this vein when, in his plea for a new "culture of repair", he emphasises ideas such as self-empowerment, community building and sustainability.[8]

What is missing in most current debates, however, is a sense of the long history of repairing things which saw ups and downs in cultures of repair and self-

5   Jackson, "Rethinking Repair".

6   Baier, Andrea et al.: "Die Welt reparieren: Eine Kunst des Zusammenmachens", in: id. (eds.): Die Welt reparieren. Open Source und Selbermachen als postkapitalistische Praxis, Bielefeld: transcript 2016, p. 34–62.

7   Grewe, Maria: "Reparieren in Gemeinschaft: Ein Fallbeispiel zum kulturellen Umgang mit materieller Endlichkeit", in: Bihrer, Andreas/Franke-Schwenk, Anja/Stein, Tine (eds.): Endlichkeit. Zur Vergänglichkeit und Begrenztheit von Mensch, Natur und Gesellschaft, Bielefeld: transcript 2016, p. 331–349.

8   Heckl, Wolfgang: Die Kultur der Reparatur, Munich: Hanser 2013.

repair, as well as the heterogeneity and interrelatedness of the actors involved. For example, the repair cultures of the interwar period, the DIY movement of the 1960s and the environmental movement of the 1970s were motivated by similar concerns for self-empowerment or sustainability. In this chapter, we will draw on examples from Western Europe and North America to highlight some important moments in the history of repair, the intrinsic links between professional and DIY repair practices, discourses on the "lifespan" of things, and changing disposal regimes.

The act of repair has always been an integral aspect of things, organisations and procedures. Viewed in abstract terms, it can be described as a reciprocal relationship between humans and their material environment which is significant in economic, social and cultural terms and ultimately inherent to each of our interactions with the things that populate our environment.[9] Reinhold Reith regards the act of repair as an "adapted technology" which offers solutions to various challenges faced by society, ranging from resource scarcity and the need to reduce waste to the way in which people spend their leisure time.[10] The act of repair does not only pertain to the level of technology but also to questions of culture and society: repairing a thing stabilises or readjusts the relationship between the thing and the user and between the thing and society,[11] and it is always possible for new meanings to be ascribed to the object in the process.[12]

---

9   Graham, Stephen/Thrift, Nigel: "Out of Order: Understanding Repair and Maintenance", in: Theory, Culture & Society 24, 3 (2007), p. 1–25; Jackson, "Rethinking Repair"; Schabacher, Gabriele: "Im Zwischenraum der Lösungen. Reparaturarbeit und Workarounds", in: ilinx – Berliner Beiträge zur Kulturwissenschaft 4 (2017), p. XIII–XXVIII.

10 Reith, Reinhold: "Reparieren: Ein Thema der Technikgeschichte?", in: Reith, Reinhold/Schmidt, Dorothea (eds.): Kleine Betriebe – Angepasste Technologie? Hoffnungen, Erfahrungen und Ernüchterungen aus sozial- und technikhistorischer Sicht, Münster et al.: Waxmann 2002, p. 139–161, p. 161.

11 Orr, Julian: Talking about Machines. An Ethnography of a Modern Job, Ithaca, NY/ London: Cornell University Press 1996; Henke, Christopher: "The Mechanics of Workplace Order: Toward a Sociology of Repair", in: Berkeley Journal of Sociology 44 (1999/2000), p. 55–81.

12 Edgerton, David: The Shock of the Old. Technology and Global History since 1900, Oxford: Oxford University Press 2007; Rosner, Daniela K./Turner, Fred: "Bühnen der Alternativ-Industrie: Reparaturkollektive und das Vermächtnis der amerikanischen Gegenkultur der 1960er Jahre", in: Krebs/Schabacher/Weber, Kulturen des Reparierens, p. 265–279.

Over the 20th century, the gradual emergence of consumer and throw-away societies meant that people – or rather those living in more affluent regions of the world – attached less and less significance to the concept of repair. This was by no means a linear process, however, with upswings in interest by different actors and at different points during this period. Mass production and consumption therefore did not directly result in a general decline of repair cultures. Rather, the history of repair over this period was characterised by multifaceted changes, relocations and shifts, with individual areas of decline but also individual areas of growth. A different history of repair cultures can be told for each different field of technology, with the added complication that certain eras such as periods of war or crisis were marked by a resurgence in repair and self-repair. In the field of commercial repair, changing cost relationships were a significant driver: in highly developed mass consumption societies, repair tasks which are difficult to standardise took a back seat in the face of ever-lower "buy-new" costs and ever-higher labour costs.[13]

In the case of large-scale production plants and infrastructure installations, however, repair and maintenance tasks have continued to play a vital and unavoidable role in extending the working life of technology for as long as possible.[14] They represent the largely invisible backbone of production, service provision and consumption opportunities,[15] yet we know surprisingly little about them, since comprehensive statistics as well as historical studies on the number of people employed in repairing or maintaining goods or selling second-hand goods, and the value these people create, are for the most part lacking. Private households also continue to repair and maintain the goods they own, but doing so is no longer the cardinal rule of housekeeping it once was; the sheer number of goods owned by most households means that most people repair only a few selected and valued items. Nevertheless, improvised repairs using quick-fix solutions such as the omnipresent duct tape are still part of our everyday interactions with the things around us, even in the "throw-away" society of the 21st century. If we step back and adopt a broader perspective, the practice of repairing old devices and continuing to use them has merely tended to shift to different parts of

---

13 Reith, "Reparieren".

14 Denis, Jérôme/Pontille, David: "Material Ordering and the Care of Things", in: Science, Technology & Human Values 40, 3 (2015), p. 338–367; Krebs, Stefan: "Memories of a Dying Industry. Sense and Identity in a British Paper Mill", in: The Senses and Society 12, 1 (2017), p. 35–52.

15 Edgerton, Shock of the Old, p. 75–102.

the world, particularly its poorer regions, which are not typically where the devices were first used.[16]

The act of repair is often held up in contrast to the practices of disposal and new acquisition, and lauded as a "better" alternative to recycling. Much is made of the fact that repairing a thing and recycling it involve two fundamentally different interactions between society and the thing itself: repairing preserves a thing's "thing-ness", whereas recycling destroys it in order to allow the thing to be reused for its materials. Between these two extremes, there are various reuse options based on disassembly of the product into its individual parts and components, some of which may potentially be incorporated at a later stage into other products as part of the repair process. Consequently, the acts of repair and recycling can figure as complementary strategies on the path towards sustainable product use and disposal. When investigating repair cultures, it is therefore important to ask questions about the "lifespan" of things, "cascades of use" for second- or third-hand things, and where exactly disposal is situated in the spectrum between throwing away and recycling.[17]

If we want to assess the relative sustainability of repair and recycling, however, we must do so in the wider context of material flows and disposal pathways. As long ago as the 1970s, environmentalists championed the ecological virtues of repairing a thing instead of producing a new thing in its place, and called for a shift towards longer product lifespans and product repairability. Whereas people had previously been motivated to repair things out of a desire to save resources and be frugal, now they were motivated to do so out of a desire to save the environment. Over the following decades, however, environmental policy focused on the recycling end of the scale rather than the repair end.[18]

---

16 Edgerton, Shock of the Old; Hahn, Hans P.: "Das 'zweite Leben' von Mobiltelefonen und Fahrrädern. Temporalität und Nutzungsweisen technischer Objekte in Westafrika", in: Krebs/Schabacher/Weber, Kulturen des Reparierens, p. 105–119; Malefakis, Alexis: "'Tansanier mögen keine unversehrten Sachen': Reparaturen und ihre Spuren an alten Schuhen in Daressalam, Tansania", in: ibid., p. 303–326.

17 Weber, Heike: "'Entschaffen': Reste und das Ausrangieren, Zerlegen und Beseitigen des Gemachten (Einleitung)", in: Technikgeschichte 81, 1 (2014), p. 1–32.

18 Weber, Heike: Reste und Recycling bis zur "grünen Wende" – Eine Stoff- und Wissensgeschichte alltäglicher Abfälle, Göttingen: Vandenhoeck & Ruprecht 2021 (forthcoming).

SHIFTING REPAIR CULTURES, THE MASSIFICATION OF THINGS AND NOVEL WASTE DISPOSAL SYSTEMS

The history of repair is typically couched in terms of a decline-and-fall story: people from the pre-industrial "society of scarcity" were forced to use strategies of repair, reworking and reuse in order to survive,[19] whereas members of the "throw-away society" have grown accustomed to using consumer goods on a more superficial or temporary basis and rarely bother to repair a thing because it is cheaper to go out and buy a new one. Furthermore, in pre-industrial society all societal classes repaired things; they were part of an "economy of makeshifts" that was characterised by high material and low labour costs. Yet we believe that it is time to rethink this decline-and-fall story, and below we propose an alternative history of repair – one which emphasises differentiation, multiple upswings and geographical displacement.

There can be little doubt that industrialisation resulted in a shift away from economies of repair to economies of manufacturing virgin products.[20] At the same time, two of the key factors in the rise of mass consumerism and consumer technologies in the 20th century were firstly that most people had access to ways of repairing consumer goods, and secondly that markets for repaired second-hand goods played an important role in the dissemination of consumer technologies. However, the ways in which people interact and intervene with things have altered over time; in particular, people have become less likely to patch and mend things and more likely to replace individual parts. The domestication of things like cars and household appliances has been reliant on emerging advice, maintenance and repair services offered by customer service departments, the development of which remains largely unresearched.[21] The vanguard technologies of the 20th-century mass consumption era were products like cars, large household appliances (fridges, ovens) and also radios and televisions, all of which required regular maintenance and repair. Yet research into the history of repair, continued use or repurposing of consumer technologies has, to date, focused solely on the car as a central symbol of status and prestige,[22] even though

19 Reith, Reinhold/Stöger, Georg: "Einleitung. Reparieren – oder die Lebensdauer der Gebrauchsgüter", in: Technikgeschichte 79, 3 (2012), p. 173–184.

20 Ibid.; Lenger, Friedrich: Sozialgeschichte der deutschen Handwerker seit 1800, Frankfurt a. M.: Suhrkamp 1988.

21 Reith/Stöger, "Einleitung. Reparieren", p. 182.

22 Harper, Douglas: Working Knowledge. Skill and Community in a Small Shop, Chicago: University of Chicago Press 1987; Borg, Kevin L.: Auto Mechanics. Technology

an estimated 110,000 people in the USA, for instance, were working in the field of radio and television repair alone in the early 1960s. But by 2006, the number of people employed in the field of "electronic home entertainment installers and repairers" had dropped to 40,000, even though the number of televisions per household had more than doubled from 1.13 to 2.6 sets, and the number of radio sets had risen by a much higher factor.[23] Shops which repaired radios in use and sold second-hand radios and televisions or other domestic electrical appliances (having purchased them, repaired them and perhaps upgraded or modified them) could be found in almost every inner-city neighbourhood in West Germany between the 1970s and the 1990s. With a handful of exceptions such as mobile phone repair shops and garment alteration services, shops like this have now all but vanished. There has been little research to date on the prevalence and evolution of these businesses, but a study carried out in 2011 found that there were just under 1,000 companies engaged in this field in Germany, employing around 36,500 people (not including car repairs, plumbing and house construction).[24]

Leaving to one side novel consumer devices such as mobile phones and printers, since the last third of the 20th century markets for household appliances in highly developed mass consumer societies have been saturated, which means that most people are purchasing replacements or multiples rather than acquiring appliances for the first time. An "inflation of things",[25] in other words a duplication and diversification of the consumer goods owned, has also been observed. The average German household, for instance, is currently estimated to own around 10,000 objects. UK households were estimated to own around ten times as many consumer electronic devices in 2010 as they did as in 1990, and the average small kitchen in the USA now contains around 1,000 things – three times

and Expertise in Twentieth-Century America, Baltimore: Johns Hopkins University Press 2007; Lucsko, David N.: Junkyards, Gearheads, and Rust. Salvaging the Automotive Past, Baltimore: Johns Hopkins University Press 2016.

23 McCollough, John: "Factors Impacting the Demand for Repair Services of Household Products: The Disappearing Repair Trades and the Throwaway Society", in: International Journal of Consumer Studies 33, 6 (2009), p. 619-626, p. 619.

24 Poppe, Erik: Reparaturpolitik in Deutschland. Zwischen Produktverschleiß und Ersatzteilnot, ed. by SUSTAINUM – Institut für zukunftsfähiges Wirtschaften, Berlin 2014, online: http://www.reparatur-revolution.de/wp-content/uploads/Studie_Reparaturpolitik-in-Deutschland-2014.pdf (accessed 21.07.2017), p. 5.

25 Heßler, Martina: "Wegwerfen. Zum Wandel des Umgangs mit Dingen", in: Zeitschrift für Erziehungswissenschaft 16, 2 (2013), p. 253–266.

as many as in 1950.[26] Repairing a thing, or paying for someone else to repair it, appears to have become a marginalised activity when viewed in the context of this huge quantitative increase in the number of new goods purchased in rich areas of the world. Whereas previously any item could be or was repaired, now only a small percentage of the things an individual owns are likely to be regarded as objects of repair. Affluent mass consumer societies no longer expect to be able to repair all the things in their households, but instead limit their repair efforts to specific things which are regarded as valuable and worthy of preservation and repair. For example, in the German consumption culture, certain household appliances such as Miele washing machines and Vorwerk vacuum cleaners are expected to have long operating lives, and consequently are also in demand in the second-hand market. But it is almost impossible to sell CRT televisions anywhere in Germany – although some are exported.[27]

The act of repairing household goods is therefore inextricably linked to underlying ownership cultures and the meaning of things. But it is also more broadly related to the prevailing disposal infrastructures which allow users to dispense with things, and to the duration of timeframes over which the things are used. These latter two considerations – the disposal of things and the duration of their use – are fundamental dimensions of the ways in which we handle things, and – much like the ownership and purchase of household things – have undergone a significant shift with the emergence of mass consumer society. Waste collection services, which first emerged in towns and cities before spreading to rural areas from the 1970s onwards, have made it easier for households to "throw things away".[28] In the intervening decades, municipalities have also been obliged to organise separate bulky waste collection services.[29] For the first time, this has allowed households to dispose of furniture, household goods and electrical appli-

---

26 Trentmann, Frank: Empire of Things. How We Became a World of Consumers, from the Fifteenth Century to the Twenty-First, London: Allen Lane 2016, p. 674.

27 Broehl-Kerner, Horst et al.: Second Life. Wiederverwendung gebrauchter Elektro- und Elektronikgeräte, edited by Umweltbundesamt, Berlin 2012, http://www.uba.de /uba-info-medien/4338.html (accessed 03.02.2017).

28 Strasser, Susan: Waste and Want. A Social History of Trash, New York: Metropolitan Books 1999; Weber, "Reste".

29 Weber, Heike: Vom Hausrat zum Sperrmüll – Sperrmüll als Phänomen der "Wegwerfgesellschaft", in: Pesch, Dorothee/Spiegel, Beate (eds.): Sparen, Verschwenden, Wiederverwenden. Vom Wert der Dinge, Oberschönenfeld: Schwäbisches Volkskundemuseum Oberschönenfeld 2017, p. 28–35. See also Weber, Heike: "Mending or Ending? Consumer Durables, Obsolescence and Practices of Reuse, Repair and Disposal in West Germany (1960s–1980s)" (this volume).

ances which would previously have been passed on to family or friends, repurposed or taken to a second-hand shop to be resold or scrapped – with a high likelihood of repair in most cases. Alternative "disposal channels" such as second-hand shops have continued to exist, but are regarded only as viable ways of getting rid of certain categories of household goods. Books and clothes, for example, both of which are prominent illustrations of the circulation of used goods in the early-modern period, still tend not to be thrown away by most households, but instead are disposed of in second-hand shops or via clothing swaps, online services such as momox, or public bookcases. After a debate lasting almost two decades on the problem of electronic waste and its potential toxicity, manufacturers are now obliged to take back and dispose of used electrical and electronic goods. They do so by recycling the devices rather than repairing them, however.[30] Beforehand – from around 1970 – large numbers of such electronic consumer appliances were regularly disposed of in normal waste bins as well as bulky waste collection systems.

A working group led by the anthropologist Nicky Gregson[31] examined in closer detail the flip side of the appropriation of things, in other words the practice of their "dispossession" and removal from the household and the meaning frameworks associated with this, and coined the term "divestment" to refer to the process. The group carried out observations of everyday life in UK households with a view to identifying the extent to which household goods were cared for, preventively maintained or repaired. In each case the answer depended on the level of skill of the members of the household, the things themselves and the question of intent, i. e. whether and how users intended to carry on using them.[32] Members of households were observed rubbing wooden furniture with beeswax and carrying out small-scale and large-scale repairs to furniture, but "quick fixes" were the dominant mode of repair.[33] Things that were no longer repaired might be kept and stored for later use or sorted out and thrown away, marking a deliberate decision by the user to end the thing's "lifespan".

---

30 Laser, Stefan: "Elektroschrott und die Abwertung von Reparaturpraktiken. Eine soziologische Erkundung des Recyclings von Elektronikgeräten in Indien und Deutschland", in: Krebs/Schabacher/Weber, Kulturen des Reparierens, p. 85–103.

31 Gregson, Nicky/Metcalfe, Alan/Crewe, Louise: "Moving Things Along: The Conduits and Practices of Divestment in Consumption", in: Transactions of the Institute of British Geographers 32, 2 (2007), p. 187–200.

32 Gregson, Nicky/Metcalfe, Alan/Crewe, Louise: "Practices of Object Maintenance and Repair: How Consumers Attend to Consumer Objects within the Home", in: Journal of Consumer Culture 9, 2 (2009), p. 248–272.

33 Ibid., p. 248.

THE "LIFESPAN" OF CONSUMER PRODUCTS

Advocates of the current repair movement are vocal about the fact that repairing a thing allows its lifespan, i.e. its time in use or in usable shape, to be extended.[34] The act of repair is viewed as a transformative practice which converts short-lived things into long-lived things, and a throw-away and resource-greedy society into a sustainable and repair-oriented one.[35] That said, there has been surprisingly little reflection on the concept that things have a "lifespan". Indeed, this very concept is, for the most part, an offshoot of the mass consumer society. In the late 19th century, the term "lifetime" was used in everyday parlance mainly to refer to people, plants and animals,[36] but it had also made initial inroads into the fields of chemistry and technology, where it was used to describe the longevity of things such as radium, wire cables, clockwork mechanisms, light bulbs or individual components of technical devices. The idea that consumer goods have a certain predictable "lifespan" came into common currency in parallel with the rise of mass production and consumption. Indeed, modelling and calculating the likely length of a product's utilisation phase represents a core strategy and an integral part of this mode of production and consumption: manufacturers' product policies are firmly based on notions of how often and for how long things will be used, and which components or designs are most likely to pay off in financial terms, as well as considerations relating to the supply, stockpiling and price of spare parts, repairability, warranties and maintenance. The figures which manufacturers have in mind are minimum requirements rather than envisaged maximum lifespans, and durability tests are therefore usually terminated when these minimum requirements are achieved. More recently, environmentally friendly product policies have been expected to take into account not only the aforementioned technical, material and economic criteria, but also environmental criteria when identifying a product's "optimum" lifespan.[37] One problem faced in this

---

34 See e. g. Wiens, Kyle: "Ich bin Reparateur. Ein Manifest für die digitale Revolution", in: Baier et al., Die Welt reparieren, p. 111–118.

35 Bertling, Jürgen/Leggewie, Claus: "Die Reparaturgesellschaft. Ein Beitrag zur Großen Transformation?", in: Baier et al., Die Welt reparieren, p. 275–286.

36 Anon.: "Lebensdauer", in: Meyers Konversations-Lexikon. Ein Nachschlagewerk des allgemeinen Wissens (1885–1892), 4th ed., vol. 10, Leipzig/Wien: Bibliographisches Institut 1888, p. 589–590; Pierer, Heinrich August: "Lebensdauer", in: Pierer's Universal-Lexikon, 4th ed., vol. 10, Altenburg: H. A. Pierer 1860, p. 190–192.

37 Rubik, Frieder/Teichert, Volker: Ökologische Produktpolitik. Von der Beseitigung von Stoffen und Materialien zur Rückgewinnung in Kreisläufen, Stuttgart: Schäffer-Poeschel 1997, p. 192.

respect is that the methodologies developed and the knowledge repositories built up during the age of mass production – in fields such as material testing, utility value research, durability tests or economic product life cycle calculations – belong to the manufacturers and have, for the most part, remained locked behind their factory gates. With this in mind, it should come as no surprise that many manufacturers stand accused of implementing planned obsolescence strategies, i. e. deliberately shortening their products' lifespans.

A similar lack of transparency can be observed with regard to product lifespans on the consumer side of the equation, even though it can safely be assumed that users have a rough idea of how long they intend to use a thing before they purchase it. Few people go out and buy mass consumer goods such as washing machines, fitted kitchens or computers in the expectation that they will pass them on to their children, but it is not merely the meaning attached to a thing which determines whether it will be repaired (and therefore whether it will continue to be used); it is also whether it can be repaired and the availabilities and costs of doing so. As labour costs have substantially increased over time, purchasing a new item is often more rational in economic terms than having the old one repaired. Furthermore, even the most cursory of glances at repair manuals from eras gone by reveals that a number of domestic maintenance procedures which were still standard in the 1970s or 1980s have since fallen into oblivion. For example, users of that time were reminded that they should regularly open up hairdryers, shavers and handheld mixers and perform certain maintenance tasks in order to keep them in working order for as long as possible.[38] Nowadays, it is largely the emotional and symbolic meaning of a thing which determines whether or not it is repaired. An OECD study published in 1982 found that consumers disposed of many household goods before they had reached the end of their useful life, a survey carried out in the USA in 1978 revealed that the majority of respondents believed that it cost too much to repair things, and studies from around the same time in Denmark and Norway highlighted the fact that over half of vacuum cleaners that were discarded were still operating correctly.[39] Moreover, putting second-hand appliances back into economic circulation has become difficult in highly developed mass consumer societies because of complex safety regulations and ever-faster innovation cycles. At the same time,

---

38 Middel, Bernd/Müller-Steinborn, Martin: Selbst Haushaltsgeräte warten und instand setzen, Munich: Compact 1989; Weber, "Mending or Ending?".

39 Organisation for Economic Co-Operation and Development (OECD): Product Durability and Product Life Extension. Their Contribution to Solid Waste Management, Paris: OECD 1982, p. 16, 31, 35 and 93.

complaints from consumers about lack of repairability, poor availability of spare parts and high repair costs have increased.[40] Nevertheless, some things appear to become so closely entwined with the user's values and routines that they are repaired even when it costs the user more than purchasing a replacement: in a study investigating repair workshops in south-west England, examples included comfortable slippers which had been broken in to the owner's liking, or a pan which the owner used to cook porridge in a known amount of time.[41]

In spite of the lack of transparency hinted at above, the term "lifespan" has become widely accepted, and "rules of thumb" are even quoted in the form of approximate expected years of service, in some cases perhaps based on the Federal Ministry of Finance's depreciation tables. For example, consumers and manufacturers, both in the 1970s and today, expect cars or washing machines to last around 10 years, although the underlying assumptions regarding usage patterns, operating methods and repair strategies are not explicitly stated by any of the stakeholders involved in negotiating this figure (manufacturers, consumers, goods testers, etc.). Specifying lifespans is purported to be a means of quantifying, in "human" years, the length of time between the date on which a thing is (first) purchased by a household and the date on which it is removed from that household. Indeed, some product lifespan calculations are based directly on methods associated with demographic statistics,[42] but generally speaking a wide variety of methods have emerged for determining the length of time over which things are used by their first owners, making it difficult to compare the figures obtained. In reality, countless devices defy such calculations, as they languish in cellars, drawers and garages in a transitional state between "no longer being used" and "ready for disposal", while others find themselves back on the side of reuse via informal channels.

Simply specifying a product's lifespan in years is an overly reductive way to talk about how things are used, how long they are used for, how they are repaired and how and why they are ultimately removed from households. This can also be seen by comparing different consumer societies. For example, the differing economic structures and frameworks of consumption in the Federal Republic

---

40 Weber, Heike: "Made to Break? – Lebensdauer, Reparierbarkeit und Obsoleszenz in der Geschichte des Massenkonsums von Technik", in: Krebs/Schabacher/Weber, Kulturen des Reparierens, p. 49–83.

41 Bond, Steven/DeSilvey, Caitlin/Ryan, James: Visible Mending. Everyday Repairs in the South West, Devon: Uniformbooks 2013.

42 For example, see the calculation methodology from a business studies perspective in Bellmann, Klaus: Langlebige Gebrauchsgüter: Ökologische Optimierung der Nutzungsdauer, Wiesbaden: Springer 1990, p. 52–78.

of Germany and the German Democratic Republic were associated with different underlying conditions for owning, repairing and using consumer goods, and for throwing them away, stockpiling them or passing them on. Socialist economies are often described in the literature as "repair societies", but there was a tendency in these economies for old devices not only to be repaired more frequently but also to be hoarded as a future resource, either to be swapped against something else or to be used for spare parts.[43] Private cars were used for three times longer than their projected lifespan of 8–10 years in the German Democratic Republic, which also resulted in greater efforts to repair and maintain them.[44] Spare car parts accounted for a significantly higher proportion of the automotive market in East Germany than in West Germany. Sometimes, however, cars had to be repaired and patched up before a new owner could even drive them because of shortcomings in production. In sum, the overall picture is more complicated than the popular perception of "long-lasting" East German designs, as illustrated for example in the 2016 documentary film *Kommen Rührgeräte in den Himmel? [Do mixers go to heaven?]*).[45]

Hans Peter Hahn recently criticised the use of "biographies of objects" as "bio-metaphors",[46] claiming that they imply a clear beginning and a distinct end to the "thing-ness" of a thing, and overlook the fact that a thing always exists in a state of interconnectedness with other things. Anthropomorphising a thing by referring to its "lifetime", or stating that it "ages" or has a second or third "life" is just as problematic, and ultimately tells us nothing about the way in which we interact with things, in other words how we use, maintain and repair them, and the practices and forms of knowledge involved in these processes. A more useful metaphor might be the idea of a cascade of use, which incorporates reuse and repurposing by a thing's new owners as well as any associated repairs and changes to the form and significance of the thing, right through to its disassembly and dismantling for spare parts – a common practice in economies of the poor – and its final disposal or even placement in a museum as a thing worthy of preservation, at which point repair becomes restoration.

43 Gerasimova, Ekaterina/Chuikina, Sofia: "The Repair Society", in: Russian Studies in History 48, 1 (2009), p. 58–74.

44 Möser, Kurt: "Thesen zum Pflegen und Reparieren in den Automobilkulturen am Beispiel der DDR", in: Technikgeschichte 79, 3 (2012), p. 207–226.

45 *Kommen Rührgeräte in den Himmel?* (2016), director: Reinhard Günzler.

46 Hahn, Hans Peter: "Dinge sind Fragmente und Assemblagen. Kritische Anmerkungen zur Metapher der ʿObjektbiographieʾ", in: Boschung, Dietrich/Kreuz, Patric-Alexander/Kienlin, Tobias (eds.): Biography of Objects. Aspekte eines kulturhistorischen Konzepts, Paderborn: Wilhelm Fink 2015, p. 11–33.

For the early modern period, researchers have investigated such cascades of use in exhaustive detail, for instance for clothes[47] and household goods[48]; these cascades led from the upper classes to the lower classes, and then ultimately to various recycling pathways when repairs were no longer possible. Over the 20th century, it became common for cascades of use to lead from initial use in a rich country to continued use in poorer regions of the world, as in the case of radios, tape recorders, used cars or second-hand clothes.[49] Repair habits and distribution routes within these recent cascades of use are often poorly documented, while at the very least the more problematic among them – e. g. old electrical and electronic equipment and electronic waste – are currently coming under much criticism.[50] The case of consumer electronics demonstrates the extent to which poorer regions of the world are used legally or illegally as a dumping ground for cast-offs from richer countries. In poor economies, such cast-offs might be put to use in new ways which differ from the original intended purpose of the consumer good. It is only recently that these repair cultures of the Global South have been studied in the fields of science and technology studies, ethnography and cultural sciences.[51]

ACTORS OF REPAIR: REPAIR AS INVISIBLE WORK AND AS AN ACT OF CONSUMPTION

A glance at the employment structure of engineers – central actors of the technical world – reveals that most engineers today are engaged in the maintenance

---

47 Fontaine, Laurence (ed.): Alternative Exchanges. Second Hand Circulations from the Sixteenth Century to the Present, New York: Berghahn Books 2008.

48 Stöger, Georg: Sekundäre Märkte? Zum Wiener und Salzburger Gebrauchtwarenhandel im 17. und 18. Jahrhundert, Wien: Verlag für Geschichte und Politik 2011.

49 Hahn, "Das 'zweite Leben'"; Malefakis, "Tansanier mögen keine".

50 Salehabadi, Djahane: "The Scramble for Digital Waste in Berlin", in: Oldenziel, Ruth/Trischler, Helmuth (eds.): Cycling and Recycling. Histories of Sustainable Practices, New York: Berghahn Books 2016, p. 202–214.

51 Edgerton, Shock of the Old; Verrips, Jojada/Meyer, Birgit: "Kwaku's Car: The Struggles and Stories of a Ghanaian Long-Distance Taxi-Driver", in: Miller, Daniel (ed.): Car Cultures: Materializing Culture, Oxford/New York: Bloomsbury 2001, p. 153–184; Jackson, Steven J./Pompe, Alex/Krieshock, Gabriel: "Repair Worlds. Maintenance, Repair, and ICT for Development in Rural Namibia", in: Proceedings of the 2012 Conference on Computer Supported Cooperative Work, ACM, New York 2012, p. 107–116.

and repair of existing things rather than the development and design of new things. David Edgerton has therefore compared the activities of engineers with the prophylactic and curative efforts of medical practitioners: "If most doctors and dentists maintain and repair human bodies, then similarly engineers are concerned with keeping things going, with diagnosis and repair of faults, as well as operations."[52] The real question, then, is why we know so little about what "repair people" do, and about the past and current significance of repair, repairability and maintenance in technical training courses and the engineering sciences.[53]

Our lack of knowledge about the people who carry out repairs and the knowledge they apply can be traced back to the fact that the act of repair, despite its omnipresent nature, is structurally invisible:[54] if a thing has been repaired skilfully – and that is the stated goal of any act of repair –, it is no longer possible to tell that it has been repaired.[55] What is more, repairs, and in particular major infrastructure repairs, often take place "behind the scenes" in order to avoid disrupting the general public.[56] For example, the hours after dusk in a European city around 1900 were not just a time for revelry, but also for repairs to rail and tram lines, tarmacked streets, and all urban infrastructures ranging from railway stations and post offices to sewage plants.[57]

Although people who make a living by repairing things form one of the largest service industries in the world, their jobs are often socially and culturally invisible and attract little prestige. According to Susan Leigh Star and Anselm Strauss, repair practices are "invisible work",[58] and their routine and everyday nature means that they are barely perceived by users, despite the fact that they are visible in principle and necessary to guarantee the operation of industrial systems, organisations and institutions.[59] "Repair people" take centre stage and be-

52 Edgerton, Shock of the Old, p. 100.

53 Reith/Stöger, "Einleitung. Reparieren".

54 Graham/Thrift, "Out of Order".

55 Schabacher, "Im Zwischenraum".

56 Denis, Jérôme/Pontille, David: "Material Ordering and the Care of Things", in: Science, Technology & Human Values 40, 3 (2015), p. 338–367.

57 Schlör, Joachim: Nachts in der großen Stadt: Paris, Berlin, London 1840-1930, Munich: Artemis und Winkler 1994.

58 Star, Susan Leigh/Strauss, Anselm: "Layers of Silence, Arenas of Voice: The Ecology of Visible and Invisible Work", in: Computer Supported Cooperative Work 8, 1 (1999), p. 9–30.

59 Schabacher, Gabriele: "Medium Infrastruktur. Trajektorien soziotechnischer Netzwerke in der ANT", in: Zeitschrift für Medien- und Kulturforschung 4, 2 (2013), p. 129–148.

come explicit mediators of events only when a malfunction or other interruption occurs.

It should be noted, however, that the social prestige of individual repair professions has varied greatly over time and between different social and historical contexts. For example, car mechanics have been looked down on in the USA since the early days of chauffeur-mechanics in the late 19th century, with complaints often voiced about unnecessary or bungled repairs, over-inflated prices and a lack of suitable trainees.[60] By way of contrast, the automotive industry in Germany has always been held in high esteem, and in the post-war decades the car repair workshop came top of the list for young men seeking apprenticeships.[61] Even in pre-industrial society, the guild system of craftsmen led to social differentiation: many craftsmen not only produced goods but also repaired them, and some, such as cobblers or tailors, specialised entirely in repair services,[62] but those outside the guild system – mostly journeymen not employed by a master craftsman – were stigmatised as "travelling journeymen" (Störer) or "false workers" (Pfuscher).[63] Generally speaking, the social status of those who earned most of their money by repairing things, including members of ethnic or religious minorities who moved around from place to place or were based in the vicinity of second-hand markets in towns and cities, was very low. As more and more consumer goods began to be produced on an industrial scale, however, the task of repairing those goods started to dominate the working lives of craftsmen in many different trades from the late 19th century onwards.[64]

Pre-industrial households habitually repaired their own goods as part of everyday life and did not pay others to do so; only higher-status households delegated repair work to craftsmen. Over the course of the 20th century, the practice of repairing one's own goods fell in and out of popularity a number of times, with marked upswings during times of crisis or war. The close ties between manufacture and repair are also evident in relation to the act of self-repair. The sewing machine, for example, was first owned as a domestic means of production in the

---

60 Borg, Auto Mechanics.

61 Krebs, Stefan: "'Dial Gauge versus Senses 1-0'. German Car Mechanics and the Introduction of New Diagnostic Equipment, 1950-1980", in: Technology and Culture 55, 2 (2014), p. 354–389, here p. 368.

62 Bernasconi, Gianenrico: "Technische Kulturen des Uhrenreparierens: Wissen, Produktion und Materialität (1700-1850)", in: Krebs/Schabacher/Weber, Kulturen des Reparierens, p. 141–162.

63 Reith/Stöger, " Einleitung. Reparieren", p. 178–182; Lenger, Sozialgeschichte.

64 Reith/Stöger, "Einleitung. Reparieren", p. 180.

late 19th century,[65] before morphing into a private mending tool over the course of the 20th century. Even as recently as the 1950s, it was still sometimes used for making clothes and sometimes for mending them, and advice manuals aimed at middle-class households contained passages on what a sewing box should contain and how a sewing machine should be used.[66]

In the 1950s and 1960s – for the first time in history – repairing things one-self became a popular pastime across all social classes and was replete with new frameworks of meaning, as evidenced by the many DIY manuals, handbooks and magazines which were published with the explicit purpose of conveying the requisite knowledge to budding DIYers. The first issue of the magazine *Selbst ist der Mann [Self Do, Self Have]* came off the presses in 1957, for example, and television programmes dedicated solely to DIY also started to appear on people's screens, the most famous of which – *Hobbythek*, aired by public broadcaster Westdeutscher Rundfunk (WDR) – ran from 1974 to 2004. While in the 19th century the act of repair was largely regarded as a subaltern practice which well-to-do households delegated to servants or third parties wherever possible, the rise of a mass consumer society was accompanied by the novel idea of DIY and repair as a crafty outlet and leisure activity for men.[67] The author of a lead story published in *Der Spiegel* in the mid-1960s noted with some bemusement that "[e]very other person seems to be busy sawing, filing, planing, drilling, painting or repairing a car."[68] Shortly afterwards, the term "Heimwerken" (DIY) appeared for the first time in German-language dictionaries to describe what was, at the time, presumed to be an almost solely male preserve.[69] For example,

---

65 Hausen, Karin: "Technischer Fortschritt und Frauenarbeit im 19. Jahrhundert. Zur Sozialgeschichte der Nähmaschine", in: Geschichte und Gesellschaft 4, 2 (1978), p. 148–169.

66 Oheim, Gertrud: Das praktische Haushaltsbuch, Gütersloh: Bertelsmann 1954, p. 315–336; Derwanz, Heike: "Zwischen Kunst, Low-Budget und Nachhaltigkeit. Kleidungsreparatur in Zeiten von Fast Fashion", in: Krebs/Schabacher/Weber, Kulturen des Reparierens, p. 197–224.

67 Gelber, Steven: "Do-it-yourself: Constructing, Repairing and Maintaining Domestic Masculinity", in: American Quarterly 49, 1 (1997), p. 66–112; Langreiter, Nikola/Löffler, Klara (eds.): Selber machen. Diskurse und Praktiken des "Do it yourself", Bielefeld: transcript 2017; Voges, Jonathan: "Selbst ist der Mann": Do-it-yourself und Heimwerken in der Bundesrepublik Deutschland, Göttingen: Wallstein 2017.

68 Anon.: "Die Axt im Koffer", in: Der Spiegel, 21 Apr. 1965, p. 47–59.

69 Voges, Jonathan: "(Arbeits-)Ethos der Freizeit? Do it yourself und Heimwerken und der Wertewandel der Arbeit", in: Dietz, Bernhard/Neuheiser, Jörg (eds.): Wertewandel in Wirtschaft und Arbeitswelt. Arbeit, Leistung und Führung in den 1970er und

the booklet *Selber reparieren – aber wie?* *[Repair it yourself – but how?]* published in 1964 provided the man of the house with a basic introduction to repairs in situations when "the pipe bursts, the tap drips or the door won't close"; his wife was entrusted only with the task of listing the repairs that needed to be done.[70] DIY has become big business in the intervening decades, and homeowners are now able to purchase cheaper and pared-down versions of tradespeople's tools and select from a huge range of paint, flooring, wallpaper paste, etc. tailored to their requirements. In many cases DIY is no longer about reducing the cost of materials or labour – a key motivation back in the 1950s and 1960s – but about filling one's leisure time and feeling the satisfaction of creating something with one's own hands, making modern DIYers similar in many ways to early car drivers and radio hams. As a result of the shift in values which took place during the 1970s and 1980s, the average (male) citizen of the Federal Republic of Germany began to define himself less on the basis of his job and more on the basis of how he spent his money and his free time, meaning that the "do-it-yourself" and "repair-it-yourself" movements came to represent creative opportunities for self-realisation – eventually also for women. Moreover, certain DIY enthusiasts in the 1970s began to perceive the act of repair as an opportunity for actively extending the lifespan of things, and even the popular television show *ARD-Ratgeber: Technik* [ARD-guide: technology] suggested looking around for second-hand components or finding spare parts at the scrapyard.[71]

But repairing by oneself also had its limits. The literature advised against tinkering with devices connected to the mains gas or electricity supply, for instance, stating that this should be left to experts on safety grounds and in view of the need for specialist tools and knowledge.[72] In the late 1970s, a debate also arose between DIYers and professional car mechanics over whether cars which the owners had repaired themselves were roadworthy. In 1977, for example, a senior employee at TÜV-Rheinland (MOT-Rhineland) suggested banning DIY car repairs, an idea that was vehemently opposed by the magazine *Selbst ist der Mann*, the mouthpiece of the German DIY scene,[73] in what was patently a clash between the commercial interests of the respective factions.

1980er Jahren in der Bundesrepublik Deutschland, Oldenburg: De Gruyter 2016, p. 73–94.

70 Fellensiek, Hans: Selber reparieren – aber wie?, Cologne: Buch und Zeit 1964, p. 39.

71 Stahel, Walter: Langlebigkeit und Materialrecycling. Strategien zur Vermeidung von Abfällen im Bereich der Produkte, Essen: Vulkan 1991, p. 223.

72 Fellensiek, Selber reparieren, p. 39.

73 Anon.: "Unser Kommentar", in: Selbst ist der Mann 21, 1 (1977), p. 11.

The alternative economic and social visions of the future which entered the public consciousness in the 1970s offered yet another potential interpretation of the act of repairing a thing oneself. As commercialised as this act may have become, it was now seen as a way to oppose the current mass consumer culture by applying one's individual productive creativity and increasing the lifespan of things through repair, thereby also bringing about social change. Repair practices were of central importance for groups such as the American counter-culturalists, and the futurist Alvin Toffler coined the term "prosumer" to describe someone who not only consumes, but actively interacts with a thing.[74] Even the pessimistic prophecies of the Club of Rome's *Limits to Growth* study made references to the act of repair as a potential way of using the earth's resources more efficiently and therefore for longer.[75]

When viewed in the context of the previous repair booms in the interwar period, during the DIY movement of the 1960s and within the various environmental and counter-cultural movements which have emerged since the 1970s, the repair movement of recent years appears somewhat less revolutionary than its representatives tend to assert. These earlier cultures of self-repair similarly embodied goals such as gaining autonomy, building communities and increasing sustainability. A common factor shared by these different repair cultures is the predominantly male connotation of the act of repairing a thing oneself, and the growing exclusion of fields of typically female pursuits such as sewing and darning. Although famous early female drivers such as Erika Mann or Ruth Landshoff-Yorck also repaired their own cars,[76] and the DIY movement was not the exclusive domain of men, gender stereotypes remained entrenched. Furthermore, for the most part the current repair movement (for all that it seeks to politicise the act of repair) also perpetuates gender-specific divisions of labour, as observed by Daniela Rosner in her ethnographic studies: women continue to outnumber men when it comes to textile repairs, whereas the opposite is true when it comes to bolting and welding.[77]

---

74 Toffler, Alvin: The Third Wave, New York: William Morrow 1980; Rosner/Turner, "Bühnen der Alternativ-Industrie".

75 Meadows, Dennis et al.: Die Grenzen des Wachstums. Bericht des Club of Rome zur Lage der Menschheit, Stuttgart: Deutsche Verlags-Anstalt 1972, p. 149.

76 Hertling, Anke: Eroberung der Männerdomäne Automobil: Die Selbstfahrerinnen Ruth Landshoff-Yorck, Erika Mann und Annemarie Schwarzenbach, Bielefeld: Aisthesis 2013.

77 Rosner, Daniela K.: "Making Citizens, Reassembling Devices. On Gender and the Development of Contemporary Public Sites of Repair in Northern California", in: Public Culture 26, 1 (2013), p. 66–70.

REPAIR WORKERS AND THEIR KNOW-HOW

The distinction made so far between expert and amateur repairers is not without its problems, even though this hierarchy is often perpetuated by the repairers themselves. The first of these problems is that the practices and tools of both groups have come to resemble each other, as illustrated by the "professionalisation" of DIY. Secondly, non-professional early adopters who could afford to purchase a particular thing such as a car, a radio or a home PC in some cases knew more and had more experience in dealing with that thing during its initial period on the market than professional repairers, who often could not afford such expensive consumer goods. The inextricable links between these two groups of repairers and their repair knowledge is also apparent from the repair literature: in the field of automotive repairs, for example, early manuals for experts and amateurs were either barely distinguishable from each other or identical.[78]

The codification of repair knowledge – in the many repair manuals, operating instructions, specialist magazines, television programmes and video tutorials – begs the question of whether this knowledge is specific to repairers and perhaps less accessible to designers, manufacturers and users.[79] On the one hand, research into the sociology and history of repair emphasises the significance of formal technical knowledge, i. e. knowing how the technology to be repaired is designed and how it operates, and of having access to structured overviews of potential defects and their symptoms as found in fault trees, for example. Only when this knowledge is possessed can the repairer carry out a systematic diagnosis of faults, using a method comparable to differential diagnosis in the medical field.[80] On the other hand, research highlights the fact that repairers need to have been repairing things for a significant length of time before they acquire the embodied knowledge necessary to handle the type of problems they encounter in everyday repair practice.[81] Douglas Harper (1987) refers to this kind of knowledge as "working knowledge". It is by no means limited to an intuitive understanding of what needs to be done, but instead is characterised by situa-

---

78 Krebs, Stefan: "'Notschrei eines Automobilisten' oder die Herausbildung des Kfz-Handwerks in Deutschland", in: Technikgeschichte 79, 3 (2012), p. 185–206.
79 Jackson, "Rethinking Repair", p. 229.
80 Krebs, Stefan/Van Drie, Melissa: "The Art of Stethoscope Use: Diagnostic Listening Practices of Medical Physicians and 'Auto Doctors'", in: Icon – Journal of the International Committee for the History of Technology 20, 2 (2014), p. 92–114; Bernasconi, "Technische Kulturen".
81 Henke, "The Mechanics", p. 70.

tional flexibility, namely the ability to select between different knowledge resources when interacting with objects to be repaired and work settings. This requires an intimate knowledge of different materials, designs and symptoms of
faults, experienced at a sensory level.[82] The diagnostic knowledge described by
Harper is also closely associated with the skills of a craftsperson, since it is difficult to separate skills such as identifying damage to a shaft using only one's fin-
gertips from those such as sensing the right torque for a screw.[83]

The unique nature of repair knowledge is also evinced by failed attempts to
standardise and automate repair tasks. Back in the 1910s, for example, Ford introduced a flat-rate system for repair tasks in the hope of tackling customers'
widespread dissatisfaction with the work done on their cars by Ford workshops,
but the idea of regulating work procedures, durations and prices for individual
repairs foundered in the face of day-to-day workshop operations which were re-
sistant to rationalisation of this kind.[84] In the late 1960s, it was Volkswagen's
turn to try rationalising its repair systems, this time with the help of automated
fault diagnosis. The Diagnosis I-System was introduced in 1968 and involved
working through a series of prescribed investigations using a test bench with
multiple testing devices. In 1971, this was followed by Volkswagen Computer
Diagnostics, which involved the completion (in some cases without human intervention) of 88 test procedures. The twofold goal of this objectivisation of fault
diagnostics was to streamline workshop operations and to restore a relationship
of trust between the workshop and its customers; ultimately, however, neither
goal could be achieved in the face of recalcitrant realities.[85] Another move in
this direction occurred in the 1970s when attempts were made to standardise radio and television repairs in parallel to the introduction of modular designs,
based on the rationale that "old-school" radio and television repair technicians
would become surplus to requirements since even an unskilled amateur would be
able to replace the modules. But these attempts also failed when confronted with
the unpredictable and transient nature of accidents, faults and wear and tear.[86]
The newly invented service computer became an indispensable tool for repairers

---

82 Ibid., p. 66–69.
83 Harper, Working Knowledge, p. 118 and 124.
84 McIntyre, Stephen: "The Failure of Fordism: Reform of the Automobile Repair Indus-
    try, 1913-1940", in: Technology and Culture 41, 2 (2000), p. 269–299.
85 Krebs, Stefan: "Diagnose nach Gehör? Die Aushandlung neuer Wissensformen in der
    Kfz-Diagnose (1950-1980)", in: Ferrum – Nachrichten aus der Eisenbibliothek 86
    (2014), p. 79–88.
86 See for example the foreword in Funkschau, special edition "Reparatur-Praxis" 15
    (1983).

working in this field, but it by no means reduced the amount of knowledge needed.

CONCLUSIONS

To sum up, the history of repair in Western consumer societies has not been a history of linear decline. Although new mass production methods and the changing relationship between labour and material costs made it more likely for things to be replaced instead of repaired, the dissemination and usage of emblematic consumer products like radios, televisions and automobiles also resulted in and relied upon new maintenance and repair services. Furthermore, second-hand markets for repaired and refurbished goods were essential for the successful diffusion of many consumer products.

We have identified three types of shift – societal, cultural and geographical – that have shaped the history of repair practices in Western Europe and the USA. Societal shifts occurred when new actor groups started to own consumer products. For example, in the interwar period members of the upper class often repaired their automobiles out of necessity, since specialist repair shops could only be found in larger cities. And when members of the working class started to own automobiles in the post-war period, they often repaired them themselves out of economic necessity. Cultural shifts occurred when the DIY movement of the 1960s and the environmental movement of the 1970s turned repairing into a rewarding leisure activity and a more sustainable means of consumption. And geographical shifts occurred when certain repair services and second-hand markets vanished in Western societies and moved to poorer regions in Eastern Europe, Africa or Asia.

Finally, we have shown that maintenance and repair practices are intricately linked with different notions of a product's "lifespan". Over time, changing disposal infrastructures and novel discarding practices have influenced the decision of whether or not to repair consumer products and made the once prevalent "stewardship of objects" obsolete.[87] However, systematic historical studies of repair, reuse and disposal practices in Western Europe and elsewhere will have to substantiate these observations.

---

87 Strasser, Waste and Want, p. 21.

MAINTAINING INFRASTRUCTURES

Repairing China's Power Grid Amidst Perpetual Warfare, 1937–1955

Ying Jia Tan

When Harbin, a city under Communist control, was plunged into darkness in September 1946, a motley crew of electricians came to its rescue and thwarted the attempts by the Nationalist regime to retake the city.[1] Chiang Kai-shek's Nationalist regime and Mao Zedong's Communist Party had turned against each other shortly after the Japanese surrender in August 1945. Northeast China, which was under Japanese control between 1932 and 1945 as the puppet state Manchukuo, became a key battleground for the Civil War. After the Soviet Red Army withdrew in March 1946, the Communists moved in to fill the power vacuum. Driven by irredentism and a desire to control industrial resources, the Nationalists deployed troops to retake the Northeast from the Communists.[2] They fired the first salvo by cutting off electrical power from the Fengman Hydroelectric Dam to Harbin. The Communists responded by mobilising workers to overhaul disused generators and restored the city's electrical power supply.

Liu Yingyuan, the worker who purportedly took charge of the repairs, became recognised as a "model labourer". He moved to the Northeast when he was eleven years old, where he started out as a packer for the British-American Tobacco Company and worked many odd jobs before becoming an apprentice in a machine shop.[3] Liu's contribution to the rehabilitation of Harbin's power infra-

1   Guanying: "Haerbin fadianchang [Harbin Power Plant]", in: Renmin ribao, 1 May 1948, p. 2.
2   For an authoritative account of the Chinese Civil War, see Westad, Odd Arne: Decisive Encounters, Stanford: Stanford University Press 2003.
3   Anon.: "Zhongguo gongren de guangrong – ji gongren chushen de changzhang Liu Yingyuan [The glory of China's workers – Remembering Liu Yingyuan, the worker who became factory director]", in: Renmin dianye 4 (Jan. 1951), p. 7–11.

structure was immortalised in the 1947 propaganda film *Minzhu dongbei* (Democratic Northeast). Speaking to reporters in 1948, the director of Harbin Power Station Xiao Changhai hailed Liu as a "labour hero", who "miraculously repaired the blades on the generators". He pointed out that when the generator broke down during the Manchukuo era, it had to be airlifted to Dalian and shipped to the Mitsubishi factory in Japan for repairs. Liu apparently completed the repairs in 28 days, all the while being "ridiculed by certain backward elements". Xiao completed his account by bringing up the heroic image from the movie of Liu flexing his muscular arms as he grasped the wrench and said, "[i]n the end, he made the untamed metal beast growl and spin obediently".[4] After distinguished service at Harbin, Liu was appointed deputy director of the power stations in Changchun and Fushun and restored power to these two Northeastern cities that suffered massive casualties during the Civil War. When the Communists came to power in 1949, Liu was transferred to Beijing, where he served as director of Shijingshan Power Station, which supplied most of the electricity for the new nation's capital.[5]

Liu was one of thousands of electrical workers and engineers swept up in the crossfire during the age of "perpetual warfare" between 1931 and 1955.[6] He nonetheless remained far from the front lines of the war between China and Japan from 1937 to 1945 and only came face-to-face with armed conflict after the Soviet invasion of Manchuria in August 1945. In comparison, China's first gen-

---

4   Guanying, "Haerbin fadianchang".
5   Anon.: "Zhongguo gongren de guangrong", p. 8.
6   The Mukden Incident of 18 September 1931 marked the beginning of armed conflict between China and Japan. The Japanese Kwantung Army orchestrated a railway bombing near Mukden (now Shenyang, Liaoning Province) and used it as a pretext to invade Northeast China. They established the client state of Manchukuo. This was quickly followed by the 1932 Battle of Shanghai, during which Shanghai became the first city to experience an air raid. On 7 July 1937, a minor skirmish escalated into the all-out Japanese invasion of China. The War of Anti-Japanese Resistance lasted until August 1945. Soon after, China descended into civil war, which led to the Communist takeover of mainland China and Nationalist retreat to Taiwan. The state of war continued even after the founding of the People's Republic of China in October 1949. This was followed by the mobilisation for the Korean War between 1950 and 1953. Although the military administration came to an end in 1953, the Chinese Communist Party continued to pursue a number of mass mobilisation campaigns that led to the militarisation of civilian life.

eration of "engineer-bureaucrats" devoted themselves to revolutionary causes and war for their entire careers.[7]

This was the case for Bao Guobao, an engineer who played an integral role in developing China's electrical industries under both the Nationalist and Communist regimes. Born in Zhongshan county in Guangdong, which happened to be the home town of the founding father of the Chinese Republic, he attended the secondary school affiliated with Jiaotong University and Tsinghua College before obtaining a Bachelor's degree in Mechanical Engineering from Cornell University in Ithaca, New York in 1922. After graduating, he taught at Zhejiang University and Jiaotong University and worked at a cotton textile mill in Zhengzhou, in Henan Province. In 1928, Bao abandoned his career in the private sector to join the National Construction Commission, an agency of the executive branch of the Nationalist government that managed state-owned power companies and regulated the private power sector across the nation. Bao oversaw the nationalisation of Nanjing Power Company in 1929. He was later appointed chief engineer of the power companies in Fuzhou and Guangzhou and reorganised their operations.[8] As will be shown later, before the war, Bao focused his efforts on replacing obsolete equipment and paid little attention to repair. After the Japanese invasion in July 1937, he began coordinating repair operations that served strategic military goals, which contributed to the accumulation of repair knowledge. Bao transferred his engineering expertise to the Communist Party when he defected after the Communist capture of Beiping (later renamed Beijing).

Drawing on published sources such as engineering journals and newspapers, this article examines how the age of perpetual warfare between 1931 and 1955 shaped the major shifts in repair culture within China's electrical industries. Before the Second Sino-Japanese War (1937–1945, in China most commonly known as the War of Anti-Japanese Resistance) in July 1937, system builders were preoccupied with replacement and not repair – a lack of expertise in repair meant that it was easier to replace faulty components. Cut off from foreign supplies after the fall of the coastal cities to Japan, Chinese engineers who retreated to Southwest China built and repaired power systems that were often blown up

---

7   This term refers to engineers employed by the government. I have chosen to refer to them as "engineer-bureaucrats" as it reflects their primary identity as members of a rationalistic bureaucracy who executed the government's policies. Their roles as engineers were secondary to their bureaucratic functions.

8   Bao, Guobao: Human resource file, 15 Oct. 1946, Aide-de-camp intelligence records, 12900000097A, Academia Historica, New Taipei City, Taiwan.

by the enemy. They scraped together any electrical equipment they could find, refurbished damaged equipment and accumulated expertise in manufacturing equipment and parts on their own through these acts of repair. "Broken world thinking", to borrow Steven Jackson's idea, was not merely a philosophical exercise for China's first generation of engineers but a normal state of affairs, as "erosion, breakdown and decay" marked the "starting points" of China's electrical industry. [9]

The Civil War (1946–1949) and the early years of the People's Republic (1949–1955) marked another turning point, which saw rank-and-file workers lay claim to the narrative of heroic repair, thereby challenging the expertise of highly educated engineers. Most of the engineers who built and maintained the power infrastructure during the Anti-Japanese War of Resistance defected to the Communists after 1949. They facilitated the transfer of knowledge across different political regimes and the codification of maintenance and repair protocol. China's first generation of electrical engineers was well aware of the inherent defects built into the power infrastructure. Power bureaus, which came under immense pressure to increase power output without installing new generators, assumed the risk of catastrophic failure by running these faulty systems close to maximum capacity. Engineers and workers addressed systematic failures with conflicting approaches. Workers chose to tinker, but engineers attempted to formalise maintenance procedures. In the end, workers came to dominate repair culture, as their actions fitted with the Communist ideology of workers seizing the means of production.

INHERITING A BROKEN WORLD

When the Nationalist government under Chiang Kai-shek established its new capital in Nanjing in 1928 after brokering peace with various warlord regimes, it inherited a broken world torn apart by years of ceaseless warfare and political turmoil. The founding father of the Chinese Republic, Sun Yat-sen, saw the construction of a national capital that was clean, efficient and offered the newest technology as the first step to achieving "national, political, and ideological uni-

---

9   Jackson, Steven J.: "Rethinking Repair", in: Gillespie, Tarleton/Boczkowski, Pablo J./ Foot, Kirsten A. (eds.): Media Technologies: Essays on Communication, Materiality, and Society, Cambridge, MA: MIT Press 2014, p. 221–239.

ty".[10] Sun died in 1925 before achieving his revolutionary goals. China's electrical industries reflected the economic reality confronting the new national government. There were 704 power stations across China, 523 of which were owned by Chinese businessmen. The average installed capacity was 394 kW, barely enough to power street lights in a small county seat and a few workshops. With average capitalisation at $105,491, these small private power plants lacked the capital to expand. Foreign capitalists dominated China's electrical power sector, as their installed capacity of 273,262 kW exceeded that of Chinese-owned power stations. There was also no national electrical power standard to speak of. Some 81.6% of China's power companies adopted a mains voltage of 220/380 V and 50 Hz in 1930, with the rest adopting 110 V and 60 Hz.[11] Fragmented and undercapitalised, these small power stations struggled to remain in business.

Nanjing's electrical infrastructure was hardly befitting for a national capital. Neglect of routine maintenance and failure to collect unpaid electricity bills caused the new capital's power company to be broken beyond repair. The Nanjing Power Company had two power stations – one in Xiaguan equipped with a 1,000 kW generator, and the other at Xihua Gate in the western part of the city with six steam engines, an apparent power rating of 840 kVA and a real power rating of 300 kW. In his report based on a January 1928 survey, former Nanjing engineer Shen Sifang admitted that "the boiler suffered from excessive corrosion, causing a thick layer of sediment to form in the boiler, thus reducing its efficiency".[12] The generator at Xihua Gate only achieved an output voltage of 1,500 V, which was half of its 3,300 V setting. Every night, it burned 29 tons of coal to generate 4,000 kWh (7.25 kg of coal per kWh). The rotor blades on the General Electric generator at Xiaguan Power Station were damaged. It generated 8,000 kWh by burning 29 tons of coal nightly, but this fuel consumption rate was still much higher than the 1 kg of coal per kWh benchmark. 60% of the electricity was lost through splicing. Although the mains voltage was 220 V, the actual voltage reading was as low as 30 to 40 V.[13] The power system in Nanjing needed a massive overhaul.

10 Musgrove, Charles: "Building a Dream: Constructing a National Capital in Nanjing, 1927-1937", in: Esherick, Joseph (ed.): Remaking the Chinese City: Modernity and National Identity, 1900 to 1950, Honolulu: University of Hawaii Press 2002, p. 139–158.

11 World Power Conference: Transactions Third World Power Conference vol. 1, Washington DC: Government Printing Office 1936, p. 117.

12 Shen, Sifang: "Zhengli shoudu dianchang gongzuo zhi yiduan [Anomalies observed in the retrofitting of Capital Power Station]", in: Gongcheng 4:2 (1929), p. 266–268.

13 Bao, Guobao: "Shoudu dianchang zhi zhengli ji kuochong [The retrofitting and expansion of the Capital Power Station]", in: Gongcheng 4:2 (1929), p. 269–271.

The replacement of obsolete generators was preferred over repair, as it delivered the desired results quickly. Bao Guobao, who assumed responsibility for the nationalisation of China's electrical industries as director of the electrical power division in the National Construction Commission, installed two small diesel generators with a generating capacity of 137 kW as a stop-gap measure. He had three reasons for doing so. One, diesel generators could be activated upon installation and offered instantaneous relief for power shortages. Two, they did not require much water, which was important since the power station was a distance away from the river. Three, they could be used as back-up power after the installation of a boiler and turbines purchased from Worthington Pump & Machinery and the Brown-Boveri Group. He also outlined plans to demarcate the service areas of both power stations to better monitor transmission voltage.[14] He offered few details on maintenance tasks such as boiler repairs and wire pole replacement; it was not that they were unimportant but they did little to resolve fundamental problems in Nanjing's power infrastructure.

Bao drew on his experience of rehabilitating Nanjing's power network when he took charge of industrial regulation and argued that the most effective way to lower costs was the replacement of obsolete equipment. Bao illustrated his point with this example: A new 1,000 kW generator, which burned 4 pounds of coal for every kWh, cost 150,000 silver dollars. The annual cost saving in coal was 57,000 silver dollars, which meant that the power company could recover the cost of the generator with three years of fuel savings.[15] For Bao, the equipment's age was not the determining factor of obsolescence. In deciding whether to replace their capital assets, power companies needed to carry out cost-benefit analyses based on comparisons with available technology.

Industrial regulation became a means to mend the "broken world" resulting from decades of mismanagement. Bao pointed out that most power companies experienced a line loss of more than 50%, which he attributed to mistakes in network design made by untrained craftsmen. Bao castigated power companies for neglecting routine inspection and condoning the theft of electrical power. He requested that private power companies submit the drawings of generators and grid layout during registration for approval. This would allow regulators to identify potential design flaws and overlapping service areas.[16]

---

14 Ibid.

15 Bao, Guobao: "Banli dianqi shiye zhe ying you zhi zhuyi [On the management of the electrical industries]", in: Dianye jikan 1:1 (1930), p. 1–5.

16 Ibid.

Private power plant operators shared Bao's concerns but saw the establishment of an electrical equipment industry as the fundamental solution to the inefficiencies of China's electrical industries. Sun Shihua, a co-founder of the Association of Private Chinese Electric Corporations, expressed his frustration with the haphazard acquisition process of Chinese power plants. Power companies lacked the capital to replace obsolete equipment. Sun noted that as a result, "old power plants were littered with abandoned machines and looked like cabinets of curiosities". Furthermore, the Chinese not only relied on imported electrical equipment but also depended on foreign suppliers for repair and maintenance.[17] Returning to Liu Yingyuan's anecdote in the introduction, the Japanese-built power company in Manchukuo had to ship the generator back to Japan for repairs when it broke down.

Chinese power plant owners cited Japan's experience to show how electrical equipment manufacturing contributed to the accumulation of repair and maintenance expertise. Sun pointed to the rapid expansion of the Tokyo Electric Company, which started operations in 1887 with a 25 kW DC generator and increased its total installed capacity to 2.23 million kW in 1925. Sun attributed its exponential growth to the emergence of a vibrant electrical equipment industry during the First World War. He estimated that dozens of Japanese electrical equipment firms held $400 million of capital and had an annual output worth $200 million. The electrical equipment industry was a foundational enterprise that offered vocational training to thousands of workers and also provided many opportunities for technical transfer through joint ventures with foreign firms. Sun himself had completed an apprenticeship in a wire manufacturing plant at General Electric in 1915, where he had learned how to work with metal and insulating materials. When he returned to China to work for the Foochow Power Company, Sun organised training sessions for ten workers, which "equipped workers with the expertise to make and repair meters, fans, motors and transformers on their own". Small-scale efforts by a local power company had a limited impact.[18] Until 1938, China's electrical industries primarily relied on imported equipment and lacked the technical expertise to conduct basic repairs.

---

17 Sun, Shihua: "Guoren ji yi chuangshe dianqi zhizao gongye [The people of our nation should quickly establish an electrical equipment manufacturing industry]", in: Dianye jikan 1:1 (1930), p. 7–11. Quotation from p. 8.
18 Ibid.

LEARNING TO REPAIR AT WAR

Sun's vision for a domestic electrical equipment industry came to fruition soon after the Japanese invasion. Plans to manufacture electrical components and military communication devices began as early as 1935. When the war against Japan broke out in July 1937, coastal cities fell under Japanese control, cutting off vital imports of raw materials and machinery. The National Resources Commission, an agency in charge of wartime industrialisation, coordinated the evacuation of China's industries inland. It had initially planned to situate the wartime industries in Xiangtan, in Hunan Province, in the Middle Yangtze region. The Japanese carried out a massacre in the Nationalist capital Nanjing in December 1937 and marched up the Yangtze River to capture the interim wartime capital in Wuhan in October 1938. The Resources Commission uprooted the industries and evacuated further inland. The engineer-bureaucrats dismantled boilers and turbines and had them transported on boats and mule carts through hundreds of miles of river valleys and mountain tracks into the mountainous southwestern provinces of Sichuan, Yunnan and Guizhou.[19] Against all odds, the Resources Commission had established a power station, copper smelting plant and electrical equipment manufacturing industry by 1939 in Yunnan's provincial capital Kunming, which would become the backbone of the wartime industrial complex. In 1938, Bao fled from Nanjing to Sichuan, where he stayed until 1944, using salvaged and refurbished components to build two power stations in Sichuan under the looming threat of Japanese air raids.

As well as restoring power supply to sustain China's war machine, repair work provided an avenue for engineers and workers to learn how to manufacture electrical equipment without foreign assistance. The Resources Commission established manufacturing facilities for wires, radio tubes and field telephones through technology transfer agreements hastily completed before 1937. As the supply of raw materials slowed to a trickle, laboratories at the electrical equipment plant developed alternative materials to relieve the shortage of spare parts. Japanese control over French Indochina after September 1940 exacerbated the problem, as it was no longer possible to transport heavy machinery through the railways between the port city of Haiphong and Kunming.[20] The Chinese now had to build generators on their own.

---

19 Tan, Ying Jia: Recharging China in War and Revolution, Ithaca: Cornell University Press 2021, p. 80–82.
20 Ibid., p. 96.

The retrofitting of a damaged generator for the Longxi River Hydropower Plant just outside the wartime capital in 1942 reveals how wartime scarcity compelled the engineer-bureaucrats to build a self-reliant industry. Zhang Chenghu, the engineer who had completed the technology transfer agreement with a British wire manufacturer, detailed the refurbishment that was completed with domestically-made components and designs drafted by Chinese engineers:

"The Longxihe Hydroelectric Plant run by the National Resources Commission purchased a 2500 kW (600 rpm) generator. The steel sheet stator was deformed and the generator coils were completely destroyed. It appointed the Electrical Works to repair and modify it into a 1700 kW, 6900 V generator, to suit the needs of the Changshou Hydroelectric Facility. Our plant took charge of the redesign, manufactured the generator coils on our own, used mica insulating material and completed the project in a year. Apart from handing the production and installation of the steel stator to the Minsheng Machine Works in Chongqing, we designed and installed every other component. The generator has been in use for nearly a year, and we have not heard of any malfunction. It is truly an accomplishment for the domestic electrical industries to complete the retrofitting of high-voltage and high-capacity power generators."[21]

The completion of the Longxi River hydropower project was a milestone for China's electrical industry. Prior to this, China's electrical equipment industry had depended on imported raw materials and made products by copying foreign designs. Years of dismantling, reassembling and repairing electrical equipment on the run had equipped the engineers with much-needed expertise to handle various types of electrical components.[22] After completing the Longxi River project, the Electrical Manufacturing Works conducted similar repairs for the damaged hydropower turbines at Yaolong Hydropower Station in Yunnan and Xiuwen Hydropower Station in Guizhou in 1943. It also applied the experience from refurbishing the hydropower generator for Longxi River to manufacture a 220 hp/750 rpm hydropower turbine for the Wanxian Hydropower Station in Sichuan in 1944.[23]

---

21 Zhang, Chenghu: "Guoying dianli jiqi shiye zhi chengzhang yu zhanwang [The growth and outlook for state-owned electrical machine industry]", in: Ziyuan weiyuanhui jikan 5:2 (1945), p. 214. Also cited in Tan, Recharging China, p. 101-102.

22 Yun, Zhen: "Wunlan lai Zhongguo dianji zhizao zhi jinbu [Improvements in electrical equipment manufacturing]", in: Diangong 1 (1943), p. 8–12.

23 Ibid, p. 11.

The electrical industry that came of age during the War of Resistance against Japan was soon caught in the crossfire of the civil war between the Nationalists and the Communists. Instead of building on their earlier achievements, engineers and workers spent most of their efforts protecting the electrical infrastructure from enemy destruction. The Communists glorified the repair work completed by rank-and-file workers. Stories of heroic repair not only proved that the workers supported the revolutionary cause of the Communists; they also showed the Communists to be competent urban administrators.

Maintenance, or the lack thereof, was a central theme of the anecdote in the opening paragraphs about Liu Yingyuan's "heroic" repairs during the Nationalist assault on Harbin in the fall of 1946. The coal-fired power plant in Harbin had fallen into disuse, so Harbin came to depend on the Fengman Hydropower Station about 252 km south of the city after 1943. In their attempt to dislodge the Communists from Harbin, Nationalist forces cut off electricity from the Fengman plant and plunged the city into darkness. The Communist newspaper *People's Daily* reported that the workers responded by working day and night to repair the disused generators and restored power one autumn evening.[24] Liu Yingyuan received the "Hero of Labour" award for leading the repair efforts.

The article about Harbin Power Plant also tells an incredulous tale of Zhang Kexing, another "labour hero" who climbed into the boiler and braved the scorching heat to replace a 200-pound moulding brick. Zhang came to Harbin at age 17 and worked at the power station for 28 years, eventually being promoted to chief of the boiler repair section. Workers recounted Zhang's superhuman feat:

"During the era of the bogus Manchukuo regime, we would have to wait for three days for the boiler to cool down before working, which would lead to a three-day power outage ... Now the workers are taking charge of the factory. None of us are willing to idle the machines that make military supplies for our soldiers on the front line. Old Hero Zhang thought of a way to quickly fix the problem. He ordered workers to remove the burning coals for the boiler and lay out thick straw mats on the furnace. The workers wrapped themselves in thick clothing and burlap sacks before crawling into the furnace, while others sprayed water to cool them down from behind. Zhang and a few workers installed a 200-pound moulding brick under these conditions."[25]

---

24 Guanying, "Haerbin fadianchang".
25 Ibid.

Stories of heroic repairs publicised through Communist propaganda laid the groundwork for workers to seize control of the means of production. Mobilisation for war forced workers to conduct swift repairs, as was the case during the War of Anti-Japanese Resistance. The spirit of self-reliance was also carried over from the earlier period. The Civil War effected a cultural change among the electrical workers. The stories of Liu and Zhang highlighted the flaws of electrical power systems and practices inherited from the Japanese. These heroic figures marked a clean break from the past, as they rallied other workers to take the lead and adopt new maintenance and repair procedures. Electrical workers came to be seen as comrades-in-arms of the People's Liberation Army. The battle-hardened engineers also threw in their lot with the Communists in the Civil War. After the founding of the People's Republic, maintenance and repair continued to be couched in militaristic language – a legacy of at least twelve years of ceaseless warfare.

REPAIRING THE NATION

The Civil War did not end with the proclamation of the founding of the People's Republic of China in October 1949. New China was quickly drawn into the Korean War in October 1950. As the electrical industry remained on a war footing, the Communists mobilised engineers and workers to guard against sabotage. The mobilisation of the masses broke down the top-down command structure of the electrical industries that had long been dominated by the educated engineering elite. The Communists created an institutional framework that welcomed input on the improvement of maintenance and repair practices from both engineers and workers. The electrical industries experienced their "let a hundred flowers bloom" moment in the 1950s, when engineers and workers proposed a vast array of ideas to improve the reliability of the power system. As we will see, workers focused on tinkering with the workflow to optimise output with limited resources. Engineers attempted to formalise maintenance procedures. This multi-faceted approach to maintenance and repair allowed the Chinese to unleash the full potential of their existing capacity.

Workers and engineers offered differing maintenance and repair solutions to improve overall efficiency and reliability. Bao Guobao, the engineer who had developed the power infrastructure for the Nationalist government before and during the war, defected to the Communists in 1949 and was appointed as the director of the Electrical Power Division under the PRC's Ministry of Fuel Resources.[26] Liu Yingyuan, the worker hero who purportedly saved Northeast China from darkness, took over as director of the Shijingshan Power Plant that supplied most of Beijing's electricity.[27] While Bao worked at the ministerial level, Liu focused on the day-to-day operations of maintaining a stable power supply to the new capital. Bao, who had served as general manager of the state-owned North Hebei Electric Company between 1945 and 1949 under the Nationalist regime, knew the vulnerabilities of the North China power grid too well. Shijingshan Power Station had been built in the 1910s and changed hands between private Chinese businessmen, Japanese regional power companies and Nationalist engineer-bureaucrats over a period of nearly four decades. As every owner installed a new generator without getting rid of the old one, the crew had to maintain generators with five different voltage settings (110, 220, 380, 3,300 and 5,200 V). It was impossible to find spare parts for the largest generator, which was built by the Japanese by imitating Swiss designs.[28] Between December 1948 and January 1949, Bao was trapped in Beiping during the siege on the city. The People's Liberation Army captured the power station moments before attacking the city. They intermittently cut off power to hundreds of thousands of inhabitants and troops, which led to the "peaceful liberation" of Beijing.[29] The trauma of war was also fresh in Liu Yingyuan's mind. In the Northeast China battlefield, Liu witnessed how the PLA pounded the enemy into submission by cutting off electricity to the city. He experienced first-hand how difficult it was to restore power to war-torn cities.

The electrical infrastructure of the new capital was no more reliable than that of the old Nationalist capital. Within the first eight months of 1950, Beijing reported 48 deaths from electrocution, while Tianjin reported 31. Residents making illegal hook-ups to the power line and thieves stealing power cables to sell them as scrap metal accounted for some of the deaths. The North China Power Bureau took responsibility for some of these accidents. It concluded that the deaths were all preventable. If the power bureau had been more diligent with its routine maintenance, accidental contact with high-voltage lines would have been

---

26 Aide-de-camp: Personnel report on Bao Guobao, circa 1950, Aide-de-camp papers, 12900000097A, Academia Historica, Xindian District of New Taipei City.

27 Anon., "Zhongguo gongren de guangrong", p. 9.

28 Xu, Ying: Dangdai Zhongguo shiye renwu zhi [Collective biography of industrialists in modern China], Shanghai: Wenhai chubanshe 1948, p. 191–202.

29 Id.: Beiping eryue weicheng ji [Two-month siege of Beiping], Beijing: Beijing chubanshe 1993.

averted.[30] Between April and July 1950, Shijingshan Power Station reported 8 to 13 serious incidents, mostly resulting from generator malfunction. An inspection team rectified seven sets of problems, including poor connections within the rotor, improper insulation of ball-bearing tubes and accumulation of dirt in fan coils and switches, and reduced the number of serious incidents to two by August 1950.[31]

Incident reports from Shijingshan Power Plant suggest that these breakdowns resulted from the overuse of existing electrical equipment and the deferment of elective maintenance. As the Ministry of Fuel Resources pursued a policy of unleashing the full potential of existing equipment, power stations across the country operated at full capacity to increase output, which left little down time for inspection and routine maintenance. It was estimated that 28.5% of the existing generating capacity had been frozen as a result of years of disrepair, and that the power output could be increased by several hundred thousand kW simply by repairing broken equipment.[32] Soviet advisers fuelled their expectations by endorsing this view. They even ridiculed their Chinese counterparts for curtailing power output out of fear of causing excessive wear and tear. China's electrical power sector did increase its output by 20% without installing new equipment,[33] but this heightened the risk of catastrophic failure. The reduction of breakdowns reported in August 1950 proved to be illusory. Problems resurfaced at the Shijingshan Power Station within a few months. Technicians skipped a major repair in November 1950 because Shijingshan Power Plant had to make up the shortfall resulting from the overhaul of Tianjin Power Plant. The operators did not remove the stator for cleaning during the minor repair in January 1951. Three months later the stator burned out, as large amounts of dirt had clogged up the generators' vents, warping the wiring insulation.[34]

---

30 Anon.: "Anquan yongdian, bimian chudian shigu [Use electricity safely, avoid accidents]", in: Renmin dianye 2 (Oct. 1950), p. 90–97.

31 Shijingshan Power Station: "Jiancha xiaozu chubu chengji shouhuo baogao [Preliminary results of the inspection group]", in: Renmin dianye 2 (Oct. 1950), p. 17–20.

32 Zhongyang ranliao gongyebu [Ministry of Fuel Resources]: Dianye jihua jiangxi ban jiangyi, Beijing: Ranliao gongyebu chubanshe 1952, p. 343.

33 Anon.: "San nian li Sulian zhuanjia gei women de bangzhu he jinhou women ruhe geng hao de xiang Sulian xuexi [Assistance rendered by the Soviet Union in the past three years and how we should learn from the Soviets from now on]", in: Renmin dianye 22 (Dec. 1952), p. 2–3.

34 Anon.: "Duiyu Shijingshan fadianchang yici tasheng zhongyao shigu de jiancha zongjie, [Conclusion of inspections following multiple serious accidents at Shijingshan Power Station]", in: Renmin dianye (May 1951), p. 9–16.

As a factory director who started out as a worker, Liu maximised power output by adjusting the workflow to improve the combustion rate of coal. When Liu won the model labourer award in September 1950, he was praised for "exterminating a huge enemy" by reducing the amount of pulverised coal that was escaping from the chimneys of Shijingshan Power Plant. Not only were workers contracting lung disease after inhaling coal dust; five to six tons of coal were escaping from the chimneys every day. The biographical essay claimed that many experts, including Soviet advisers, had attempted to solve the problem, but only Liu had set his mind to it.[35] Liu resorted to offering of 20,000 *jin* of millet to any worker who fixed the problem. The workers implemented a joint suggestion from two workers and the Soviet advisers to patch up a ventilation gap in the coal pulveriser, but that only solved half of the problem. Two other workers modified the design of the coal conveyor to stabilise the feed and improve the burn rate. These modifications saved 2,000 tons of coal each year, which reduced the cost of power generation.

Workflow adjustments allowed Liu to identify safety lapses. On 5 June 1950 two workers overfilled the steam generators and caused water to seep into the turbine. Liu made an example of the negligent workers and highlighted the importance of "exterminating" accidents. He demoted the workers at fault and rewarded those who detected problems. His biographical sketch does not indicate whether he took any measures to rectify systematic mechanical faults causing generator breakdowns.

While workers tinkered with the workflow, engineers focused on identifying weak links accumulated as a result of the rapid succession of regime changes. A Japanese-made generator installed during the war accounted for most of the breakdowns. It was jointly manufactured by Hitachi and Mitsubishi in the 1940s based on imitations of AEG's products, but the windings for the rotors were made with thinner material and inadequately stretched, so they broke easily, taking out the stator's winding and silicon steel plates at the same time.[36] Generator malfunction was not the most common problem. According to the Ministry of Fuel Resources, 32.6% of all breakdowns across the nation in the first quarter of 1951 were caused by faulty transformers, 70% of these because of damage to the windings. Referencing Soviet performance benchmarks, the engineers noted that transformers operating at a normal temperature of 85° C had an expected shelf

---

35 Anon., "Zhongguo gongren de guangrong", p. 8.
36 Zhongyang ranliao gongyebu [Ministry of Fuel Resources]: Sulian dianye zhuanjia baogao dierji [Soviet electrical experts' report], Beijing: Ranliao gongyebu chubanshe 1951, p. 178–182.

life of 20 to 25 years, but as the insulation wore off, the operating temperature increased to 150° C, causing the transformer to burn out within a few years.[37] Transformers with worn-out insulation were largely built during the War of Anti-Japanese Resistance, when both Japan and China faced material shortages.

The engineers lamented that workers carried out maintenance tasks without understanding fundamental principles and ended up damaging the equipment they wanted to maintain. In his speech during the safety inspection meeting in September 1950, Bao related an anecdote of a foreign technician who overfilled the oil in the steel bearings, causing the power station's water pump to break down. He filled the steel bearings with the best oil, but the oil deteriorated quickly and had to be replaced every two weeks. Upon further inspection, the engineers realised that the oil protected the bearings from corrosion but did not serve as a lubricant. Overfilling the bearings with oil increased friction, which led to excessive wear and tear.[38] The mechanic at fault was purportedly a foreigner who had worked in the plant before 1949. The story reflected the realities of the electrical industries during the early Communist takeover. Bao's remarks were aimed at the ad-hoc inspection committees that covered up underlying problems rather than fully resolving them.

The engineers felt compelled to formalise maintenance protocols and eliminate systematic errors that caused catastrophic failure. As early as January 1951, Bao Guobao appointed an eight-person taskforce to compile a "Manual for Inspection and Repair of Electrical Equipment". Two of its members, Wang Ping-yang and Yu En-ying, were among the first batch of engineers sent to the Tennessee Valley Authority for advanced training during World War Two. Wang specialised in high-voltage transmission lines and participated in the construction of the 154 kV power line from the Apalachia Dam. Yu En-ying began his training in the design department but later branched out to focus on the "commercial and management side" of electrical power distribution.[39] During their

---

37 Ibid, p. 180.

38 Bao, Guobao: "Anquan jiancha zhong ji ge ying zhuyi de wenti – jiu yue ershi ri zai anquan jiancha gongzuo huibao shang de jianghua [A few problems to take note of during safety inspections – speech at the safety inspection meeting on September 1950]", in: Renmin dianye 3 (Oct. 1950), p. 14–15.

39 Cheng, Yu-feng/Cheng, Yu-huang (eds.): Ziyuan weiyuanhui jishu renyuan fu mei shixi shiliao [Archives on the National Resources Commission Technicians' Training in the United States], Taipei: Academia Historica 1990, p. 122. The chapter in the "Maintenance Manual" on protecting the electrical grid from lightning strikes is largely based on the notes of Sun Yun-suan, a TVA trainee who was sent to Taiwan and

advanced training at the TVA, Wang and Yu submitted monthly training reports to their supervisors at the National Resources Commission. These reports provided the raw material for the technical manual. Draft chapters were circulated in the bi-weekly *Renmin dianye* (People's Electrical Industries) journal. Engineers and technicians across the country were encouraged to share their experience and offer suggestions for revision. The engineers employed techniques of mass mobilisation to transform the compilation of the technical manual into a collective enterprise.

The engineers recognised maintenance lapses with generators and transformers as the most pressing problems and published these draft chapters first. The first chapter went straight to the underlying cause of generator burnouts and then spelled out instructions for the routine cleaning of the rotor and the stator. The manual specified that "The dust on the coils had to be blown with compressed air from 1.7 to 2.5 kg/cm^3, or removed with a vacuum cleaner then wiped off with a cotton cloth. The grease on the coil may be removed with carbon tetrachloride, but not too much, as the substance does not evaporate easily and can damage the insulation after getting absorbed into it".[40] Overall, the manual adopted a comprehensive approach towards maintenance and repair. Every section adhered to a fixed format with an inspection checklist, maintenance procedures, diagnostic methods for common problems and a troubleshooting guide. There were also pictorial guides that showed readers diagrams of commonly-used tools and detailed every single step of the maintenance procedures. The publication of the technical manual also coincided with establishment of an incident reporting system by the Ministry of Fuel Resources. The magnitude of potential hazards became clear after the second National Electrical Power Conference in 1951. In his April 1951 report, Bao's deputy director Zhang Bin noted that seven regional power networks had identified a total of 95,500 problems but were only able to rectify 8.32% of them.[41]

The published sources offer no information on the effectiveness of the formalisation of maintenance practices. Power stations across China, including those in North China, continued to report high incidences of serious accidents. In

---

later rose through the ranks to become Taiwan's premier, published in this edited volume (p. 640–647).

40 Anon.: "Dianqi shebei jianxiu bixie [Draft of Handbook for the Maintenance of Electrical Facilities]", in: Renmin dianye 7 (Jun. 1951), p. 3–39. Quotation from p. 5.

41 Zhang, Bin: "Guanyu anquan gongdian he jishu bao'an gongzuo de jidian wenti [A few problems regarding the safe provision of electrical power and technical security]", in: Renmin dianye 7 (Jun. 1951), p. 1–8.

the nine months of 1953, Qingdao Power Station in Shandong reported 20 outages, with 18 of them happening in the month of July. Lightning struck the transmission cables during the rainy season in summer. Since lightning arresters were in short supply, the power bureau did not install them on some of the transmission towers and was therefore unable to "avoid a power outage whenever there was a thunderstorm".[42] Incident reports recognised the need for regular maintenance but acknowledged that there were lapses in quality control, so much so that it was "better to leave things unrepaired".[43] The special issue on the "increase production, practise economy" campaign offered some clues about the electrical power industry exceeding quotas despite the high frequency of power outages. Of the seven reports, four of them detailed how power plant workers increased output by reducing fuel wastage during transportation and combustion. The only report that attributed increased output to stringent maintenance came from the Beijing Power Bureau. 120 power line workers from East China went to Beijing to certify the overhaul of Beijing's power network ahead of National Day in 1953. The remaining two articles were about power conservation efforts by the cotton mills in Shanghai and paper manufacturing in Shandong. Put simply, workflow improvements implemented by rank-and-file workers offered quick fixes that translated to observable results. Workers, not engineers, took credit for productivity gains.

CONCLUSIONS

Material scarcity during the War of Anti-Japanese Resistance (1937–1945) led to an involuntary transition from replacement to repair in China. The wartime economic blockade disrupted the supply of electrical parts to the country's fledgling electrical industry. The state-run power companies under the Nationalist regime in Southwest China were particularly hard hit. Their engineers and technicians accumulated considerable experience in repairing battle damage, as they dismantled electrical equipment from coastal cities and transported it over land, rehabilitating the power infrastructure of wartime Chongqing and Kunming after damaging air raids. Those who performed heroic repairs for the Nationalist regime during the War of Anti-Japanese Resistance defected to the Communists after

---

42 Zhong, Shufeng/Yu, Longhai: "Qingdao dianye ju guanli boru, shigu buduan fasheng [Weaknesses in technical management in the Qingdao Power Bureau; Accidents keep happening]", in: Renmin dianye 13 (Oct. 1953), p. 28.

43 Ibid.

1949. Their expertise in repair and maintenance allowed the Communists to take over the electrical infrastructure with minimal disruption.

The workers' approach of tinkering with the workflow became more visible than the systems analyses performed by engineers, since it fit into the Communist Party's overall strategy of improving productivity through mass mobilisation. Elaborating on Lenin's formula "Communism is Soviet government plus the electrification of the whole country", Zhang Bin, head of the Electric Bureau in the Ministry of Fuel Resources, noted that "The transition towards communism requires the electrification of the whole country, as it requires the entire national economy to move towards large-scale production".[44] Workflow adjustments aimed at improving the electrical industry's fuel efficiency involved the participation of many more workers than specialised maintenance tasks. Repair and maintenance became a means of imposing discipline on the vast industrial workforce.

During the first five-year plan (1953–1957), the electrical industry increased electricity supply not primarily through the addition of new generating capacity but by spreading power demand evenly throughout the day with the readjustment of production schedules. As a result, the maintenance of a safe and efficient power system became the collective responsibility of all workers. The expertise developed through decades of tinkering and repair under conditions of material scarcity would culminate in the November 1958 mass electrification campaign, during which the Ministry of Electrical Industries called on the entire population to build improvised power plants to relieve chronic power shortages following the launch of the Great Leap Forward.

---

44 Zhang, Bin: "Dianli gongye zai fazhan guomin jingji zhong de zuoyong [The role of electrification in the development of the national economy]", in: Renmin dianye 15 (Aug. 1955), p. 4.

Changing Perceptions of Repair and Maintenance:

Reflections from the Oral History of the

Electricity Supply Industry in the UK

Thomas Lean

In 2014 I filmed an oral history interview in a coal-fired power station in central England. Built in the 1960s, the station was still operational and its original turbo generators still provided close to their design output nearly half a century from their installation (though aptly, one of them was offline for repairs that day). Standing in the control room I was struck by the juxtaposition of old and new required to keep this cathedral of power running. Along one side of the room were modern computerised control desks, desktop PCs and large flat display screens monitoring every aspect of the plant's operation. On the other side was a wall of switches, dials and glowing light bulb indicators, housed in grey metal cabinets proudly carrying the logo of an electronics company that had been defunct since 1968.[1]

1   I am deeply grateful to all those who were interviewed for An Oral History of the Electricity Supply Industry (henceforth OHESI); for the generous philanthropy of the project funders, Hodson and Ludmila Thornber, for enabling the project to happen; for the advice of the project advisory committee, Sir John Baker, David Jefferies, Stephen Littlechild and Leslie Hannah; for comments on drafts by Sally Horrocks, Heike Weber and Stefan Krebs; and for the support of all the archive staff at National Life Stories who care for the OHESI archive, including Rob Perks, Mary Stewart, Elspeth Millar, Emily Hewitt, Haley Moyse, Eleanor Lowe, Camile Johnson, Charlie Morgan and Cai Parry-Jones. An earlier version of this paper was presented at the Histories of Technology's Persistence: Repair, Reuse and Disposal workshop in 2018 and I am grateful for the suggestions of other delegates.

As this anecdote suggests, electricity supply industries (ESI) are rich systems in which to observe the persistence of technology. The longevity of ESI engineering is a testament not only to the robust quality of original design but also to the care it receives over decades of operation. Yet infrastructure is largely invisible except when it fails, and the activities of the many people who are responsible for its ongoing care over decades often go unnoticed. Graham and Thrift argue that "the processes of maintenance and repair that keep modern societies going … have been neglected by nearly all commentators as somehow beneath their notice".[2] Accordingly, whilst there has been some attention to how the electrical system was operated during times of trouble, particularly in Frank Ledger and Howard Sallis' *Crisis Management in the Power Industry*,[3] the repair and maintenance of the ESI has not been given a great deal of attention in its history in Britain. For example, the subject is mentioned only briefly in Leslie Hannah's classic history of the post-war industry until the early 1980s, and even less in Dieter Helm's account of the more recent development of the sector.[4] The focus in these and other works has tended to be on business history, technical development, economics, organisation, politics and the growth of the industry.

Historians of technology have long stressed the importance of adopting a systematic approach to understanding the development of electricity industries.[5] A similar perspective is required to understand their maintenance and repair. The integrated nature of electricity networks means that unchecked local faults can lead to widespread cascading failures. Maintenance work has to be carefully programmed to allow items of equipment to be switched out and isolated, both for the safety of workers but also to protect the integrity of the system as a whole. The British electricity network is a 24/7 operation and, as electricity cannot be stored, supply and demand on electricity networks need to be kept con-

2 Graham, Stephen/Thrift, Nigel: "Out of Order: Understanding Repair and Maintenance", in: Theory, Culture & Society 24, 3 (2007), p. 1–25, here p. 1.

3 Ledger, Frank/Sallis, Howard: Crisis Management in the Power Industry, London: Routledge 1995. The major focus here is on political crisis, notably the miners' strikes of the 1970s and 1980s. Both authors served in the ESI and Frank Ledger was amongst those interviewed for OHESI.

4 Hannah, Leslie: Engineers, Managers and Politicians, London and Basingstoke: Macmillan 1982; Helm, Dieter: Energy, the State, and the Market, Oxford: Oxford University Press 2003.

5 Hughes, Thomas P.: "The evolution of large technological systems", in: Bijker, Wiebe E./Hughes, Thomas P./Pinch, Trevor (eds.): The Social Construction of Technological Systems. New Directions in the Sociology and History of Technology, Cambridge, MA/London: MIT Press 1987, p. 51–82.

stantly in balance by the coordinated actions of individuals across the industry. Overall, this system presents an excellent example of the sort of "broken world" discussed by Steven Jackson, a piece of infrastructure only kept going by the constant attention of maintenance and repair workers whose efforts are generally hidden except when parts of the system fail.[6]

In this paper I use oral history interviews to explore the history of repair and maintenance from the perspective of electricity supply industry workers. The interviews were collected between 2013 and 2017 as part of *An Oral History of the Electricity Supply Industry in the UK* (OHESI), led by National Life Stories (NLS) at the British Library.[7] Established in 1987, National Life Stories' mission is "to record first-hand experiences of as wide a cross-section of society as possible, to preserve the recordings, to make them publicly available and encourage their use".[8] This has been principally achieved through a series of oral history projects that have focused on recording and archiving the life stories of people who worked in particular industries or sectors of society.

NLS interviews are typically semi-structured, relying on a biographical outline, open-ended questions and a conversational interview style, and recorded over several sessions. Eight- to fifteen-hour interviews are quite normal, giving interviewees the chance to reflect at length on their lives within a wider social history framework and to recall daily life and developments in the areas they worked in. OHESI collected over 500 hours of recordings with more than 60 people connected to the electricity industry. Most of them were professional en-

---

6   Jackson, Steven J.: "Rethinking Repair", in: Gillespie, Tarleton/Boczkowski, Pablo J./ Foot, Kirsten A. (eds.): Media Technologies: Essays on Communication, Materiality, and Society, Cambridge, MA: MIT Press 2014, p. 221–239.

7   Further details on project background are available in the project scoping study: Horrocks, Sally/Lean, Thomas: An Oral History of the Electricity Supply Industry. Scoping Study for Proposed National Life Stories Project. National Life Stories, 2011, https://www.bl.uk/britishlibrary/~/media/subjects%20images/oral%20history/oral%20hi story%20and%20nls%20documents/nls_electricityindustryscopingstudy.pdf (accessed 29.10.2018). Further details on project implementation, including interviewee selection, are available in the end of project report: Lean, Thomas: An Oral History of the Electricity Supply Industry. Final Report. National Life Stories, 2017, https://www. bl.uk/britishlibrary/~/media/Oral%20History-Electrical%20Supply%2023102017.pdf (accessed 29.10.2018).

8   "British Library Projects: National Life Stories", https://www.bl.uk/projects/national-life-stories (accessed 17.06.2019).

gineers with long careers in the sector, but the collection also included accounts from some industrial staff, managers and others connected to the industry.[9]

Repair and maintenance were not the specific focus of the interviews, but the issues came up in conversation with several interviewees as they described the work that they had once done and commented on how the industry had changed over their careers. As such, the corpus cannot be read as a continuous dataset but more as a collection of snapshots from across the industry's past; a series of vignettes and reflections on repair and maintenance from a wide range of people. This includes people who worked both in power stations and on the transmission and distribution system. Naturally, the viewpoints in the interviews are subjective, but as oral historians have long argued, this very subjective quality can allow oral history interviews to reveal salient truths. Oral sources, as Alessandro Portelli points out, have a "different credibility". They can show us not just what people did but how they understood the meaning of these activities both then and as they retell them now.[10] OHESI interviews contain three principal sorts of information about repair and maintenance: stories and anecdotes about experiences of repair and maintenance jobs, descriptions of the work activities involved, and reflections on the meanings of these activities and how these meanings could change over time. I illustrate this essay largely with the third category, interviewees' subjective views, but present them in analysis and context based on other information that emerged in interviews.

My particular focus in this essay is on how maintenance and repair were changed by the privatisation of the British electricity industry, i. e. how practices of repair as well as the perception and meaning of repair changed. Through most of the careers of those interviewed, the ESI was owned and operated by the state, after nationalisation by Clement Attlee's left-wing Labour Government in 1948. For most of the era of nationalisation, the Central Electricity Generating Board (CEGB) ran the power stations and nationwide high-voltage transmission system through which electricity was dispatched to a dozen Area Electricity Boards in different regions of the country, which in turn distributed it through low-voltage networks to customers.[11] After the 1979 election of Margaret Thatcher's right-

---

9   All interviews cited in this essay were conducted by Thomas Lean for OHESI, and are generally available online at http://sounds.bl.uk/Oral-history/Industry-water-steel-and-energy (accessed 12.03.2019).

10 Portelli, Alessandro: "What Makes Oral History Different?", in: Perks, Robert/ Thompson, Alistair (eds.): The Oral History Reader (2nd edition), London: Routledge 2003, p. 67.

11 The industry was organised differently in Scotland and Northern Ireland, where the functions of the CEGB and Area Boards in England and Wales were more vertically

wing Conservative Government, bent on free-market reforms, the ESI was privatised over the 1980s and 1990s. The CEGB was split up into competing generating companies and an independent National Grid company was established, and the Area Boards became regional electricity companies. The whole industry became increasingly fragmented as more companies became involved in the years that followed. Interviewees often present privatisation as something of a millenarian event, heralding a period of great cultural change in the industry. As the head of one privatised power company recalled of the change, "I used to say to the people at Powergen [one of the new post-privatisation generating companies]: 'If you don't know what to do ... think of what you would do in the CEGB and then do the opposite.' ... I intended them to say: 'Well if we did it in the CEGB like that, but now we're a totally different culture ... then how do we do it now?'"[12] As I discuss, there are many aspects to this cultural shift, but it can be very briefly summarised as a move away from a centrally planned and engineering-focused service industry to a competitive energy market of agile businesses that were more driven by financial profit and loss.

The privatisation and liberalisation of infrastructure systems and public services is a subject that has elicited considerable comment, much of it critical. Graham and Thrift, for example, suggest that the application of neoliberal ideologies to electrical systems has been a, "complete rebuttal of the subtle cultures of repair which actually allow complex technosocial systems like electricity to work",[13] contributing to their degradation and increasing unreliability. I do not make any value judgment here as to whether maintenance and repair arrangements were better or worse either side of privatisation. As I note later, figures from the industry regulator seem to suggest modest improvements in system reliability after privatisation. However, excusing the occasional exceptional event, the lights stayed on fairly reliably before privatisation too. Yet to listen to the impressions of those who worked in the industry through this period, it is evident that, to them, something changed in how this was achieved as the culture of the industry shifted. Various studies have highlighted how repair and maintenance activities are seen differently by the particular cultures they are part of, such as those espousing ideas of self-sufficiency, creativity or environmentalism. My aim in this essay is to explore how the different cultures of the nationalised and

integrated into single organisations. While OHESI interviews suggest some variations in culture between regions, there seems to be a common commitment to public service and engineering.

12 Interview with Edmund Wallis, OHESI, C1495/26, 2014 15.

13 Graham/Thrift, "Out of Order", p. 14.

privatised industries contributed to a change in how repair and maintenance were approached and conceptualised, though the end result, the lights staying on, has remained constant.

THE NATIONALISED ESI: ENGINEERS AND PUBLIC SERVICE

It is generally accepted that the nationalised ESI was a largely engineering-led organisation. Dieter Helm, for example, notes the "cult of the engineer" as being a major theme of the nationalised ESI, which concentrated on "ever bigger and more technical investments ... free from the constraints of competitive markets".[14] As I discuss in detail elsewhere, respect for the "heroism of practical knowledge" was deeply engrained in the culture of the nationalised ESI.[15] Most of its senior managers were originally trained as engineers, generally after following an apprenticeship with periods of hands-on work that encouraged an appreciation of engineering and craft skill. As one senior manager recalled of his apprenticeship alongside craftsmen in the 1950s: "I remember being amazed at the sheer skill of people ... who could make their tools sing and that was not with any book learning".[16] Industrial workers, including the skilled craftsmen who actually did most of the maintenance work under the direction of engineers, were often trained by the industry itself through its training schools and thorough apprenticeships. "You felt as though you were getting quality training and you were working for a quality outfit", recalled one foreman of his apprenticeship. This contributed to an organisational culture where engineering, high standards and practical ability were much respected.[17]

The ESI also imbued its staff with a sense of public service and the importance to society of maintaining electricity supplies.[18] As one interviewee put it: "You've got to do the best you can to keep the lights on and serve the public".[19] In an ethnographic study of workplace maintenance staff, Christopher Henke argues that repair of faults is essential for maintaining the normal order of

---

14 Helm, Energy, State and Market, p. 16.
15 Lean, Thomas: "The Life Electric: Oral History and Composure in the Electricity Supply Industry", in: Oral History 46 (2018), p. 56–58.
16 Interview with Granville Camsey, OHESI, C1495/09, 2013–14.
17 Interview with David Williamson, OHESI, C1495/43, 2015–16.
18 Lean, "The Life Electric", p. 61–62.
19 Interview with Frank Ledger, OHESI, C1495/01, 2012–2013.

the workplace.[20] Similarly, "keeping the lights on" helped to preserve the normal order of society, but if supplies were threatened by faults then repair staff had a duty to restore normality as swiftly as possible. "There was an ethos there that you looked after people …", recalled one distribution engineer, "[y]our job was, if they're off supply, to get them back as quickly as possible".[21]

These two aspects of ESI culture are both evident in stories of repairing faults or keeping the system going under pressure. These sometimes take on the character of "war stories", struggles to keep the engineering of the system going and serve the public despite challenges such as failing technology or horrendous weather. Julian Orr notes in an ethnography of photocopy technicians that such stories are useful in transferring knowledge and building a sense of professional identity.[22] This latter function seems particularly important in the oral history interviews, contributing to building an identity of self-sacrificing public servant and able engineer.[23] Electricity was a 24-hour industry and repair staff could be called on at any hour. Distribution engineers, for instance, served shifts at home on standby, awaiting the night-time telephone call that would summon them to attend to a fault. As one senior manager recalled of his time as a distribution engineer early in his career:

"I think the most exciting part of some of it really is, and I think partly it's still the same today although it's much more controlled, is you get rung up about 2 o'clock in the morning on the telephone, you get told that customers [are] off supply, people ringing up all over the place. And *you* have to go out, and *you* have to find the problem, and *you* have to solve it, and *you* have to repair it. Nobody else. That I guess was the challenge."[24]

There could also be an element of public appreciation too, expressed by interviewees in anecdotes about customers being grateful for their electricity supply being returned. Another retired distribution engineer, recalling various ways that faults and their repair had inconvenienced the public, reflected that people were generally, though not totally, understanding of the efforts made to get their supply back on: "Some people were less pleasant and there were one or two who

---

20 Henke, Christopher: "The Mechanics of Workplace Order: Toward a Sociology of Repair", in: Berkeley Journal of Sociology 44 (1999/2000), p. 55–81, here p. 44.

21 Interview with Michael Butterfield, OHESI, C1495/39, 2015.

22 Orr, Julian: Talking about Machines: An Ethnography of a Modern Job, Ithaca/London. Cornell University Press 1997, ch. 8.

23 See Lean, "The Life Electric", p. 56–62.

24 Interview with Bryan Townsend, OHESI, C1495/05, 2013.

were aggressive, but on the whole people were remarkably supportive for what you were trying to do".[25] On the other hand, if the system failed for too long, public displeasure could be quite unpleasant. In the early 1970s, for instance, when power cuts became a daily feature of life as striking miners cut off coal supplies to power stations, interviewees recall arguments with their neighbours over why their electricity was off and hiding electricity board insignia on their uniforms whilst out in public. It has been argued that the work of maintainers only becomes visible when systems fail. While a handful of interviewees express a certain sadness that the public were generally ignorant of the daily efforts that went into keeping the system going, this lack of awareness is perhaps a tribute to how well it was maintained and operated on the whole: an invisible system was a working system.

THE PRIVATISATION OF THE ESI:
CREATING A BUSINESS-LED INDUSTRY

Before nationalisation in 1948, the British ESI was a patchwork of over 300 different municipal and private undertakings. In the years of post-war austerity, power cuts and voltage reductions were a regular feature of daily life. Nationalisation facilitated a dramatic development of the ESI's capacity and technology and by the 1980s it had evolved into a generally reliable and robust system. The capability of the system as a whole to meet exceptional strain was demonstrated during the 1984–1985 miners' strike, the most bitter industrial dispute in recent British history. Striking miners brought the industry to its knees in the early 1970s, but a decade later the lights stayed on as the ESI withstood "the greatest challenge … of the industry's history".[26] However, while the ESI was unquestionably meeting its mission to keep the lights on, nationalised industries as a whole were under criticism for their perceived inefficiencies. Privatisation was a major policy plank of the Conservative Government over the 1980s, partly for ideological reasons, but it was also presented as a way of reforming nationalised industries. The memoirs of political figures driving privatisation, for example, go into some detail about the problems of the nationalised ESI, which they identify as muddled central planning, inefficiency and high costs.[27] This reminds us

---

25 Interview with Alison Simpson, OHESI, C1495/55, 2016.
26 Ledger and Sallis, Crisis Management, p. 293.
27 See, for example, the perspectives of two energy ministers in the 1980s: Lawson, Nigel: The View from No. 11, London: Transworld 1992, chs. 12, 15 and 16; Parkinson,

that determining whether a technical system is broken depends not only on whether its technology fulfils its function, but also whether it is perceived as working by the standards of its political-economic context.

Many senior industry figures describe privatisation not as a political issue but as a radical way of reforming problems they perceived in the state-run ESI, such as over-staffing, inefficiency, government interference and muddled or "gold-plated" technical development.[28] As one Area Electricity Board chairman put it: "I think the industry had stagnated, it had ossified ... monolithic, too big, too ungainly ... it was just ready for a massive change".[29] In the memories of industry figures in favour of it, privatisation appears almost as a way of repairing the industry's problems and improving efficiency. On the other hand, there were those who saw little wrong with the existing arrangements. As a senior manager from the CEGB recalled: "I couldn't really believe anybody could do the job better than we were doing".[30]

Perhaps the biggest difference in post-privatisation companies has been the decline in influence of engineers. Dieter Helm comments that "whereas engineers led the nationalised industries, their role was greatly reduced in the 1980s and 1990s".[31] Privatisation saw a decline in the status of engineering in the ESI, as management positions, formerly dominated by trained electrical engineers, became more open to figures with expertise in accountancy, human resources and other general management experience. Clearly engineering was still vital, but as financial targets assumed far greater significance than before, the industry embarked on a wave of cost-cutting measures that would have a number of effects on maintenance and repair. Examples noted by interviewees include the reduction of extensive power station stockpiles of spare parts, the closure of maintenance depots, with staff expected to travel further to fix problems, and new information and communication technology being used to increase maintenance staff utilisation rates.

Whilst most interviewees regard the public service mission of the industry as continuing after privatisation, a few also note a change in emphasis as financial issues and targets became more explicitly significant. As one recently retired dis-

Cecil: Right at the Centre, London: George Weidenfeld & Nicholson Limited 1992, ch. 13.

28 Lean, "The Life Electric", p. 62–65.

29 Interview with Bryan Townsend.

30 Interview with Frank Ledger.

31 Helm, Energy, State and Market, p. 16.

tribution engineer who worked on the system before and after privatisation re-marked:

"I think that was the attitude for a long time: that we were a public service ... I suspect that people now have as much satisfaction, but certainly I think that concept of 'we're do-ing what we're doing for the public' as a nationalised industry had, I'm not sure. The em-phasis of management changed from 'we're doing what we're doing for the public' to 'we're making profits for the company'. And while people still do and get satisfaction from restoring customers on and dealing with customers there's always that thing: and yes we have to make money. And sort of: we want to get customers on fast, because it's good to get customers on fast, but because if we don't get them on fast we won't meet our tar-gets or ... we'll end up having to pay them whatever the money, the financial cost, is."[32]

The shift in philosophy evident in the quote above is a clear example of how regulation and the need of the regulator to measure progress came to change how staff carried out their jobs, and how this approach could clash with the values the industry had previously instilled. The industry regulator monitored and imposed targets for reliability, repair and customer service, and required financial com-pensation to customers if their supplies were off for too long. Public service was now something with targets and financial implications for failure, and there was perhaps a subtle shift in industry emphasis from "public service" to "customer service".

The need to save money also led to changes in the people who were actually doing maintenance and repair work. The nationalised industry, while making some use of contractors, relied heavily on its own large force of directly em-ployed maintenance staff, who it had often raised from its own apprenticeship schemes. In 1983, for example, the ESI as a whole employed some 83,026 in-dustrial workers directly, not to mention some 25,675 engineers, out of a total workforce of some 158,025.[33] This technical staff of over 108,000 included op-erations and design staff, but it seems a safe assumption that the majority were probably working in repair and maintenance. Post-privatisation staff numbers steeply declined as part of an effort to cut costs. Whilst it is hard to find exact numbers, as of 2011 the whole sector was reckoned to employ just 87,000 peop-

---

32 Interview with Alison Simpson.
33 The Electricity Council, Industrial Relations Department (ed.): Digest of the Electrici-ty Supply Industry's Manpower Statistics 1983, London 1983.

le.[34] Clearly there are fewer maintainers than before. Post privatisation, electricity supply companies also increasingly contracted out maintenance and repair work they would once have largely done using in-house staff to external companies to reduce direct staff costs.[35] The result was a more flexible main-tenance workforce, but also a smaller and cheaper one.

In the interests of flexibility and efficiency in the time after privatisation, maintenance staff were also called upon to broaden their skills and knowledge. One post-privatisation electricity company chairman recalled the introduction of "multi-disciplinary, well-trained staff, they would get paid more money as a result of being able to go in and do the one job, and they would be fitters and jointers all wrapped up into one."[36] Specialist maintenance departments for electrical, instrument and mechanical aspects were merged together, with fewer staff overall. Industrial workers were up-skilled to take responsibilities previously held by more highly trained – and highly paid – engineers. Large reductions in staffing and the amalgamation of different sorts of work were partly facilitated by a decline in the power of trade unions. In the nationalised ESI, unions were strong; many managers recall being frustrated by unions' opposition to new ways of working and saw privatisation as a means of changing this. As the electricity company chairman quoted above recalled of the 1990s, it was "quite obvious that we could actually improve the efficiency of our overall labour forces ... because of the unionisation that had not been something that we'd really tackled with as much energy or as vigorously as we would do if we were independent companies".[37]

DECLINE OR DIFFERENCE: POST-PRIVATISATION CHANGE

The nationalised ESI had not been uninventive in developing new maintenance and repair technologies, but given its large workforce it did not have the same motivations to introduce labour-saving innovations as private companies did. With a greater imperative to cut costs and a smaller and more flexible workforce,

---

34 Anon.: "UK energy sector: facts and figures", in: Daily Telegraph, 13 Jul. 2011, https: //www.telegraph.co.uk/finance/newsbysector/energy/8633356/UK-energy-sector-facts-and-figures.html (accessed 05.04.2019).

35 This is a commonly mentioned point in several interviews from OHESI.

36 Interview with David Jefferies, OHESI, C1495/23, 2014 15.

37 Ibid.

private companies had more need and freedom to introduce new technology and methods to speed up maintenance work, and sometimes to allow them to be done by workers without extensive skills training. One interviewee who had worked as a cable jointer in the 1980s before privatisation spoke early in his interview about the complex work of connecting cables together "in a hole in the ground on a rubber mat wearing a pair of Wellingtons [rubber boots]":

"You'd got two ladles … you'd configure the cable in such a way so you could pour molten metal over the bound conductors and catch the residue in the larger of the two ladles. Often this would be done in a confined space … always uncomfortable, unpleasant. There'd be molten metal involved. There'd be the fumes from the flux and all the dirt, it might also be raining … the hole might be filling up with water … and then you would wrap a rag around the plasticised metal and just wipe it … until you'd got a nice smooth surface and that would eventually set hard and become a joint … then you would get rolls of … impregnated paper oil tape and wrap them continuously … tie that off with a piece of oil-impregnated string … whilst you were pouring this molten metal over the exposed conductor all the metal and the ladles were live at 240 volts."[38]

As this description suggests, there was often a considerable level of craft skill required to do repair and maintenance work in the ESI, not to mention some hardship too. Later the same interviewee explained how new methods of joining high voltage cables together, using snap-together parts and heat-shrunk materials, meant that such work could be done far more quickly and with a lower level of craft expertise:

"Many years previously it was a very, very highly skilled job and only a very small number of staff were capable of that level of craftsmanship, but as technology moved on, and as I got into the section I was looking after, it quickly became apparent that … the technological side of it, the physical side of jointing, it had been dumbed down quite a bit, you know there wasn't the same level of physical skills involved."[39]

The contrast between these two quotes and the evident appreciation of practical skill and sadness at its demise that they convey provide a glimpse into how staff perceived the changing culture of the industry. Much of the drive to cut costs after privatisation came from the industry regulator, established after privatesation, whose remit included imposing targets for expenditure and customer service on

---

38 Interview with David Williamson.
39 Ibid.

the privatised companies. The imposition of regulatory targets and constraints on spending was intended to promote business efficiency and value for money. Levels of capital investment, including maintenance and equipment renewals, were thus influenced by the regulator. "Mainly it's the engineering that drives it", reflected one regional electricity company technical director, but decisions to renew equipment in the private sector were based on a mix of financial and engineering issues, more explicitly than in the past:

"If you've told the regulator you're going to spend a certain amount of money, the regulator is not an engineering organisation, it's an economic organisation so they tend to go by the money... and if you stick to the original plan you'd underspend, because of the difficulty of doing things, then spending money on some of the easier things that you might have done next year, but doing it this year, to hit your spend target, that's a financial-driven investment plan ... you're moving something from next year to this year, if you've got a 50 year life what difference does it make? But you can't do that every year, you've absolutely got to unwind that, otherwise you build up a legacy of problems. So year on year, you're always happy to move things around, but you've got to be very careful you don't keep putting some of the hard stuff off forever. Otherwise it never gets done."[40]

Perhaps the clearest example of how the private sector has invested less in the industry than the state may have done can be seen in power stations. Other than a short rush to build new, highly efficient, gas-fired power stations in the 1990s, the privatised industry has preferred to "sweat the assets", to get as much use as possible out of its existing technology rather than investing in new technology. Although many coal-fired stations were closed in rationalisation measures in the 1990s, some have been kept in operation far beyond their expected service lives. Indeed, the oldest stations in Britain in 2019 were commissioned in 1968, and are now a decade beyond the 40-year service lives expected when designed.

From the perspective of those actually doing the maintenance work there seems to have been a decline in standards of care compared to the engineering-led nationalised industry. They remark, for instance, on how private companies seem to have cut back on how much maintenance was done, as one power station control room operator recalled:

"The periods between maintenance became longer ... they lengthened the period of time on routine plant maintenance so they would run longer 'cos it didn't cost as much, took the chance on it breaking down. And in some instances some of the plant was, because it

---

40 Interview with Mike Kay, OHESI, C1495/44, 2015–16.

was longer between maintenance, it became more leaky, but they still felt it was cost effective to put up with that ... so the place did deteriorate somewhat."[41]

A distribution foreman recalled further stories of cost-cutting from out on the networks:

"Look at the condition of the substations, most of them look semi-derelict, they're overgrown, the paint's peeling off ... you can imagine the state of some of the switch gear ... We always used to maintain substations on a six-year period, and then we came across new terminology ... 'We're going to sweat the assets': We were going to make them last a lot longer before we did anything to them ... One particular company ... they had to develop all kinds of weird, elaborate techniques of operating the switchgear by lanyards with pulleys ... tugging on a rope to operate a switch because that substation hadn't been maintained ..."[42]

As I have noted, the engineering and workmanship standards of the nationalised industry were high. The apparent decline in this standard after privatisation was something that some maintenance staff, brought up in a culture of high standards, found distasteful, as the above examples suggest. However, other interviewees, particularly engineers turned managers, suggest that the engineering standards of the nationalised industry were perhaps higher than they really needed to be. As one distribution company technical director noted: "Although we were always cost conscious [in the state-run industry], I think we were very capable of doing lots of activities that didn't add much value ... It's certainly true that in the past we used to over-maintain things".[43]

Judging from the interviewees' accounts, the privatised industry seems to have performed less maintenance than the nationalised industry. However, we cannot simply characterise this as neglect. Rather there seems to be a different approach to this sort of work; it was not about applying uniformly high standards and maintaining by rote, but rather taking an approach influenced by risk management, cost-benefit analysis and greater use of information:

"We had certain standards to keep up and from the electrical point of view we were discarding work, you know, instead of doing something once a year we'd make it once every two years type of thing. A lot hard decision-making was involved there, where you'd look

---

41 Interview with Kevan Gee, OHESI, C1495/36, 2015.
42 Interview with David Williamson.
43 Interview with Mike Kay.

at past history ... If it failed, it depends on where it was, if it [was an] ancillary plant you can say if it fails it's not going to have too much detriment ... if you had like a main breaker and that failed then the unit was off and it'd be costing megabucks per hour then you don't even risk it. And that's what they call risk management, a lot of it goes on now ... When it was CEGB, CEGB had one standard and that was top notch, it was Rolls-Royce standard all the time, without a doubt. The time element didn't really matter. It was a case of once you took a piece of plant out and it was maintained then it wouldn't break down, that's the sort of level it was at. Whereas perhaps nowadays instead of something taking five hours, it might take two, because you've cut your work content down a bit to not do certain items, you've cut corners but there's a risk element of it, where you know what you're doing."[44]

Information technology also brought about changes. Maintenance and renewal in the nationalised industry seems to have been a largely planned and routine activity; things were maintained when they were due to be maintained. However, with new ways of understanding data about how assets had been used, diagnostic tools and remote monitoring, privatised companies have developed a greater understanding of the condition of their engineering assets and when they need maintenance or replacement than in the past. As one National Grid technical manager explained:

"We gradually moved away from nameplate lives to start to look at the environment the assets had worked in, the duty [they] had done ... and started to do diagnostics associated with examining the piece of kit and deciding actually, you know, it may have been there 40 years but it looks pretty new, why replace it? ... Starting to be forensic, more scientific about when stuff needed to be replaced has been a feature right through from 1990 ... You are just more able to understand the condition the asset's in and to replace it at exactly the right moment, rather than thinking: 'Well, 40 years is up. It's time for a new one.'"[45]

Some have argued that worldwide the liberalisation of energy industries has led to a neglect of electricity networks, which has contributed to a decline in their standard of service.[46] However, it is quite difficult to judge how the changes privatisation clearly brought to maintenance and repair have affected the reliability

---

44 Interview with Brian Moore, OHESI, C1495/35, 2015.
45 Interview with Nick Winser, OHESI, C1495/32, 2015–16.
46 See, for example, Timothy, Luke: "Power Loss or Blackout: The Electricity Network Collapse of August 2003 in North America", in: Graham, Stephen (ed.): Disrupted Cities. When Infrastructure Fails, Abingdon: Routledge 2009.

of electricity supply in Britain – not least because they do not seem to have made a great deal of discernible difference on the outside. To some extent, this may reflect a legacy of generous margins in equipment installed by the state industry, with its robust approach to engineering. Whether new equipment installed by the privatised industry proves as long-lasting and resilient remains to be seen. Presently there are concerns for the future of electricity generation, as older power stations are decommissioned and less consistent renewable energy sources replace them. However, other than occasional severe weather events and local issues, the reliability of the privatised electricity system as a whole does not yet seem to have been a major issue affecting electricity supply in Britain.

According to figures from the electricity regulator, from the point of view of the consumer, not much changed regarding reliability of supply in the first decade after privatisation. In 1998 the regulator found that "there have not been major changes in the security of supply for any company since Vesting".[47] Indeed, overall the story has been one of gradual slight improvement in overall security of supply, albeit with problems in a few companies from time to time. Nationally, in 1990/1991, the first year of privatisation, there were 111 interruptions per 100 customers.[48] By 1997/1998 this had fallen to 88 and the trend of gradual improvement largely continued afterwards.[49] Yet as we have seen, this outward picture of stability conceals an enormous amount of change within the ESI itself and, in the memories of those who experienced this change, a different philosophy of maintenance. There is a lesson here perhaps not just for electricity supply industries but for infrastructure more generally: as a changing context of politics, technology and business affects the culture of the organisations responsible for caring for our infrastructure, many things have to change for them to remain outwardly the same.

---

47 Office of Electricity Regulation (ed.): Report on Distribution and Transmission System Performance 1997/98, Nov. 1998, p. 6. https://www.ofgem.gov.uk/sites/default/files/docs/1998/11/2556-dats97-98_1.pdf (accessed 18.09.2019).

48 Office of Electricity Regulation (ed.): Review of Public Electricity Suppliers 1998-2000, Distribution Price Control Review: Consultation Paper, May 1999, p. 63. https://www.ofgem.gov.uk/ofgem-publications/78992/review-oes-1998-2000-dpcr.pdf (accessed 18.09.2019).

49 Office of Gas and Electricity Markets (ed.): Ensuring a Secure and Reliable Gas and Electricity Supply 2013, p. 2. https://www.ofgem.gov.uk/ofgem-publications/59150/ensuring-secure-and-reliable-gas-and-electricity-supply-pdf (accessed 18.09.2019).

CONCLUSIONS

Oral history offers a valuable route for revealing the hidden world of the repairers and maintainers of infrastructure. Yet whilst these stories of late-night callouts and descriptions of fixing things provide a glimpse of the activities that kept the lights on, the subjective assessments that accompany them are perhaps even more revealing. They suggest that maintenance and repair in the state-run industry, with its emphasis on high standards and craftsmanship, were influenced as much by the industry's "cult of the engineer" ethos as by other aspects of its activities. They hint too at the value of public service in the identity of the repairers and maintainers who kept the lights on for society. As I have shown through analysis of these testimonies, interviewees seem to have viewed privatisation as bringing about a different philosophy of repair and maintenance. While engineering and public service have remained features, interviewees seem to regard the privatised industry as having become much more led by business concerns, finance and targets than in the past. Uniformly high engineering standards have given way to more flexibility and efficiency and perhaps lower engineering standards. The lights have stayed on, but in these subjective assessments the issue is perhaps not whether technical problems have happened or not. Rather, interviewees' perceptions of these changed, lower or more flexible maintenance standards sometimes seem in conflict with the values the state-run ESI instilled in its workforce, its emphasis on "Rolls-Royce" engineering and public service. On a personal level, for some repairers and maintainers the values of the privatised industry seem to have conflicted with the personal identities that they had built up in the state-run ESI.[50]

Whilst the conclusions of this essay must be limited by the subjective nature of its source base, the changes it attributes to privatisation and the motivations behind this shift warrant further examination. In a technological sense, the nationalised ESI worked well; it kept the lights on. However, as Britain's national politics changed in the 1980s, it was not seen to be working well in a political-economic sense. In this context, while much of the technology remained in place, the "repair" of the industry as a whole was affected by privatisation changing the organisational culture and the practices around that technology. This process suggests that histories of maintenance and repair need to think in broader terms of business organisation, politics and different perceptions, of

---

50 For a detailed discussion of how industry managers negotiated this change in industry identity, see Lean, "The Life Electric", p. 62–65.

whether or not systems are broken, rather than just whether the technology itself functions as it should.

Business as Usual:

Telephone Repair and Maintenance
at the Bell Telephone Company of Canada

Jan Hadlaw

This chapter examines histories of repair and maintenance in the early years of North American telephony, when the telephone industry was dominated by two monopolistic and closely associated telephone companies – the American Telephone & Telegraph Company in the United States (henceforth AT&T) and the Bell Telephone Company of Canada (henceforth Bell Canada or Bell).[1] It focuses in particular on the repair and maintenance practices of Bell Canada and its manufacturing arm, Northern Electric, between the 1880s and the 1930s.

Established in 1880 with capital and expertise supplied by AT&T, Bell Canada quickly acquired all but one of the telephone patents in Canada and a monopoly charter from the Canadian government. By 1887, the challenge of Canada's vast

---

1 AT&T was also commonly called "Bell" in the United States. For the purposes of clarity, I distinguish between the US telephone company and the Canadian telephone company by referring to them as AT&T and Bell or Bell Canada, respectively. In instances when I refer to these two companies together, I will use the term "Bell companies". The relationship between AT&T and Bell Canada was defined by a complex share-holding arrangement, in which AT&T's percentage declined from 48.8% in 1885 to 0% in 1975. Unlike other regional Bell operating companies, Bell Canada retained its ownership and had its own research and development labs and its own manufacturing branch, Northern Electric Company Limited. Bell Canada's relationship with Northern Electric paralleled AT&T's relationship with its equipment manufacturer, Western Electric in the United States, and the two manufacturing companies similarly shared information and technical components as well until the mid-1960s. See Rens, Jean-Guy: The Invisible Empire, Montréal: McGill-Queen's Press 2001, p. 65 and 217–218.

geography and the reluctance of Canadian investors prompted the company's directors to consolidate operations in Québec and Ontario, which were Canada's most populous provinces and home to its largest and most prosperous urban centres, Montréal and Toronto. Headquartered in Montréal, Bell Canada was never part of the (American) Bell System, but it did function as a Bell operating company, and the two companies were closely linked, freely sharing technical information and business strategies, until the mid-1960s.

Bell operating companies, including Bell Canada, were unique among their competitors in recognising the important role of repair and maintenance in the development and proper functioning of telephone networks (fig. 1). Unfortunately, the routines and rationales of repair operations were not carefully documented during the period prior to the introduction of Scientific Management techniques to Bell's operations in the late 1920s. Information on what was repaired and how, as well as who was responsible for repairing, is difficult to ascertain from company records.[2] Similarly, and perhaps not coincidentally, the practices and routines of repair have received less than their due of attention by historians of telephony. Like other scholars interested in histories of repair, I am attracted to the avenues of investigation that the study of repair and maintenance opens up.[3] For historians of technology, shifting the focus away from histories

---

2   Unlike North American independent telephone companies, Bell operating companies allocated a percentage of revenues for maintenance and repair and drew attention to this fact in their publicity and promotions. A Bell operating company spokesman noted that a "telephone plant deteriorates rapidly – more rapidly, perhaps than the mechanical equipment used by any other modern industry", that any telephone company that did not make such investments was failing both its subscribers and its shareholders. A pamphlet produced by another Bell operating company suggested that Bell companies allotted 8% of the cost value of their networks to maintenance and repair. See MacDougall, Robert: The People's Network: The Political Economy of the Telephone in the Golden Age, Philadelphia: University of Pennsylvania Press 2014, p. 144–145.

3   For example: Denis, Jérôme/Pontille, David: "Material Ordering and the Care of Things", in: Science, Technology, & Human Values 40, 3 (2015), p. 338–367; Jackson, Steven J.: "Rethinking Repair", in: Gillespie, Tarleton/Boczkowski, Pablo J./Foot, Kirsten A. (eds.): Media Technologies. Essays on Communication, Materiality, and Society, Cambridge, MA/London: MIT Press 2014, p. 221–239; Krebs, Stefan: "Dial Gauge versus Senses 1–0: German Car Mechanics and the Introduction of New Diagnostic Equipment, 1950–1980", in: Technology and Culture 55, 2 (2014), p. 354–389; Lucsko, David N.: Junkyards, Gearheads, and Rust: Salvaging the Automotive Past, Baltimore: Johns Hopkins University Press 2016; Strasser, Susan: Waste and Want: A Social History of Trash, New York: MacMillan 2000.

of innovation and discovery allows closer attention to be paid to what David Edgerton has called "technology-in-use", to the materiality of technological arte-facts under study, and to the economic and social aspects of the practices and re-lations of labour involved in the activities of maintenance and repair.[4] It also, certainly in the case of the telephone, makes the interconnection between techno-logical systems and human networks of workers and users more explicit and vis-ible.

Figure 1. In this advertisement Bell Canada boasts about the scale of its repair operations and offers instructions on how to care for telephones (Montreal Gazette, 3 May 1917, BCA, File: Newspaper Clippings 1910–1918).

The repair and maintenance of a "technology-in-use" as vast as a telephone sys-tem is necessarily ongoing and occurs at many sites. The local loops, trunks, and switching offices, and the millions of kilometres of wire that link these compo-nents, are all potential sites of breakdown and failure. Likewise, telephone users

---

4   Edgerton, David: "From Innovation to Use: Ten Eclectic Theses on the Historiog-raphy of Technology", in: History and Technology 16, 2 (1999), p. 111–136.

also constitute potential sites of technological breakdown, particularly during the period under study here. For this reason, I focus on the repair and maintenance of telephone subscriber equipment, in other words the technological instruments people used to make a telephone call: the telephone sets and private branch exchanges (PBXs) that were installed in residences and offices in Canada's urban centres. While Bell operating companies famously documented and quantified almost every aspect of their operations after the adoption of Scientific Management methods in the 1920s, information on Bell Canada's early repair and maintenance practices was more difficult to find and had to be gleaned from references in contracts, meeting minutes, and internal correspondence. The most fortuitous and useful source of information was a set of oral interviews conducted with retired repair workers who worked in Bell Canada's Telephone Contract Division at different times between 1907 and 1983. These interviews provide rare insight into the work of repairing telephones and the working lives of telephone repairmen that, like telephone use itself, are typically considered too mundane to warrant documentation. By looking at conditions on the ground through the accounts of those Bell employees most directly involved in the maintenance and repair of telephone equipment, this chapter contributes to the project proposed by Andrew Russell and Lee Vinsel to "knit together stories and anecdotes – microhistories, if you will – into an overarching narrative of maintenance".[5] This narrative is valuable, in and of itself, for what it tells us about technologies-in-use and the myriad ways that technologies were integrated into modern societies.

Furthermore, I suggest that understanding the historical role of maintenance and repair in the telephone industry is especially important today, for the insights it offers into the persistence of technology in the pre-liberalised telephone industry and beyond. Owning and leasing its telephones, and making significant investments in the maintenance, repair, and reuse of its equipment, allowed Bell to expand its networks and, most importantly, to safeguard its monopoly, during times when telephones were in short supply or when technological components malfunctioned. In turn, it can be argued that Bell Canada's monopoly status provided the conditions that made the integration of maintenance and repair in its operations both possible and logical. It is notable, though, that AT&T and Bell Canada chose to expand and rationalise their maintenance and repair operations in the mid-1920s, the same period that the majority of North American manufacturers embraced planned obsolescence. Key to making sense of this decision was

---

5   Russell, Andrew L./Vinsel, Lee: "After Innovation, Turn to Maintenance", in: Technology and Culture 59, 1 (2018), p. 1–25, here p. 7.

the associated companies' conception of the telephone as a technology and a utility rather than a commodity. This distinction made the maintenance and repair of telephone technology, or its persistence, a central concern of telephone operations. The telephone industry's valorisation of technology's persistence is instructive, especially when considered in the context of the present environmental crisis, to which the mass disposal of telecommunications devices contributes. Documenting the essential role played by repair and reuse in this modern technological industry helps to demonstrate that there was nothing natural or inevitable about 20th-century capitalism's embrace of obsolescence and disposability, and perhaps offers a means to imagine new ways of thinking about technological devices and their users.

MAINTAINING THE "TELEPHONE PLANT"

During the first 100 years of telephone service, Bell Canada, like AT&T, owned not only its systems' cables and wires, switchboards and central exchanges, and the public telephones found in city streets and inside stores, hotels, train and bus stations; it also owned the private telephones used by its residential and business subscribers. Altogether this technological apparatus – from the cables and wires to the telephones installed in homes and offices – made up what the Bell companies called their "telephone plant", a single operational entity that they owned, managed and maintained.

The affiliated Bell companies' conception of the "telephone plant" – with the telephone set imagined as just one component of the larger telephone system, and not a product, appliance or commodity – is significant for historians of technology because it distinguishes them from other utility companies and helps to explain why the repair and maintenance of subscriber sets was critical to the operations of the Bell telephone companies. Most utility companies – electric and gas companies, for example – provided services but did not manufacture or own the lamps, stoves and furnaces that their subscribers used. Since the consumers of these utilities purchased and owned their lamps, stoves and furnaces, it followed that it was their responsibility to maintain, repair or replace them as they saw fit. The condition of any one customer's technical devices was of little concern to a utility company because it did not affect the operation of its larger system. This was not the case with the telephone system, however, because subscribers using inferior telephone sets would experience inferior service and so would anyone they were calling.

Retaining ownership of and responsibility for the telephone set, and leasing it to Bell subscribers, was an early and strategic decision. It gave the Bell companies a degree of control over the technical and political problems associated with maintaining both the telephone system and their monopoly status. Bell Canada was granted a monopoly in 1882 by the federal government by arguing that its telephone service was "a work for the greater advantage of Canada".[6] Government support for Bell's monopoly rested on the company's argument that single ownership was the only way to guarantee quality of service across the entire telephone system, and so subscribers complaining of inferior service could threaten Bell's monopoly. By providing them with telephone sets that met its technical standards, Bell was able to ensure the quality of service and meet its government-imposed mandate. Ownership of "the entire plant" also allowed Bell to control the pace of technological change, and thereby to avoid the early obsolescence of equipment by introducing new technologies or components only when it was financially advantageous to do so. The leasing system also allowed Bell to monitor and maintain the quality of its telephone sets. Under Bell's leasing arrangements, when subscribers cancelled their service or moved, their telephone set would be disconnected and removed. All disconnected telephones were returned to Bell's storerooms, where they were inspected and repaired or reconditioned. Worn parts would be replaced, technical components that had become obsolete would be updated and the telephone's appearance would be refreshed if necessary, its casing polished and new telephone cords fitted.

As a consequence of the political and technological arrangements of both AT&T and Bell Canada, their telephones were manufactured to distinctive criteria. Unlike manufacturers of lamps, stoves and furnaces, for whom consumer preferences and popular style trends were key factors in company decision-making, the affiliated Bell companies' overriding concern was that the technological instruments and components that made up its telephone plant, including subscriber sets, should be long-lived, durable and repairable.

INTEGRATING MAINTENANCE AND REPAIR

At the turn of the century, the North American telephone network was still a work in progress and the telephone industry was far less organised than it would

---

6   Armstrong, Christopher/Nelles, H. V.: Monopoly's Moment: The Organization and Regulation of Canadian Utilities, 1830-1930. Philadelphia: Temple University Press 1986, p. 72.

become by the mid-1920s. Bell's 1885 attempt to stave off competition by buying up independent competitors and attempting to build a long-distance network left it chronically undercapitalised, just as demand for telephone service was beginning to grow. The rising demand for telephone service was a mixed blessing for Bell. More subscribers meant more revenue but also more expenditure on new central offices and new equipment. At this time, Bell purchased its telephones from a number of manufacturers (including Western Electric, the manufacturing arm of its American counterpart AT&T), but telephone manufacturers were unable to produce enough telephones and switchboards to satisfy demand.[7] Bell's repair operations provided it with the ability to (almost) keep up with demand, as well as giving it a degree of control over the costs and quality of equipment. The great demand for telephone service – and the difficulty meeting it – was a concern that reached the highest offices at Bell. In 1882, Bell's vice president wrote to the manager of the Ontario division enquiring about unused equipment that appeared in his inventory report: "Are these out of order? If so, you can ship them here [to Montréal], and we will repair them. We wish to put all this old material in order, that we may be able to utilize it as much as possible."[8] Any telephones that could be repaired were, so that they could be put into service.

The minutes of a meeting of Bell's senior officers and agents held in May 1887 demonstrated both the company's understanding of the importance of maintenance and repair and the nascent state of the telephone industry at that moment. Decisions about who would do the work of inspection, maintenance and repair and how it was to be done were still unresolved. The need to establish a system of inspection, whereby an inspector would be employed to visit central exchanges and inspect the equipment on a regular basis, was identified, but there was little agreement on whether this person should also be expected to do repair work. It was initially proposed that inspectors should only carry out inspections – that is, they should be separate from repair work, with Bell's vice president and managing director noting that inspectors should not be put in the position of inspecting repairs that they had made in the past. One district manager countered

---

7 The number of Bell subscribers in Québec and Ontario grew eight-fold between 1906 and 1929, from 95,145 to 761,456. See Mussio, Laurence B.: Becoming Bell, Montréal: Bell Canada Enterprises 2005, p. 40. The number of telephones in service in Canada grew rapidly between 1920 and 1930, from 856,000 in 1920 to 1,403,000 in 1930. Armstrong/Nelles: Monopoly's Moment, p. 295.

8 Memo from C. F. Sise, Vice President, to Hugh C. Baker, Manager, 12 Jan. 1882. Bell Canada Archives (henceforth BCA), Repair Service, Subject file: Correspondence 1882.

that Bell should purposely seek out agents capable of doing both their own inspections and repairs, while another proposed that the inspector "should be a sort of travelling repairer". Yet another agent suggested that, in smaller offices, telephone operators might also be trained in repair work.[9] The meeting yielded no conclusive agreement about how the company should organise inspection and repair work, thus leaving managers and agents to continue making their own judgements about how best to maintain their telephone equipment.

Nevertheless, evidence of Bell's desire to develop systems of quality control over its operations can be seen in these discussions.[10] An important step in this direction took place in 1895, when Bell established its own manufacturing subsidiary, Northern Electric, to build and repair telephones for its system as well as for independent telephone companies. In September 1912, Bell Canada and Northern Electric formalised their relationship by signing an agreement that established what would come to be known as the "Telephone Contract Division". It designated Northern Electric as Bell Canada's agent responsible for procuring "by manufacture, purchase or otherwise ... any apparatus, supplies, or materials" that the telephone company might require. The term "otherwise" undoubtedly referred to repair work: "Operating a local repair and emergency shop" was notable on the list of the new division's responsibilities.[11] The growing demand for telephone service exceeded the production capacity of Northern Electric and that of independent manufacturers as well. By repairing and reusing telephones that it already owned, Bell was able to fill more subscribers' orders and continue to expand its network.

---

9   Minutes of a meeting of the officers and principal agents of the Bell Telephone Co. of Canada, Montréal, 16 May 1887, BCA, B-26606, p. 4–5.

10 Paul J. Miranti describes how the Bell System's efforts to "certify the reliability of its equipment and to provide economical service" between 1877 and 1929 provided the "learning" that resulted in the gradual development of an "organizational structure for coordinating and controlling quality assurance activities at both the staff and line levels and between the corporate elements of the Bell System". Miranti, Paul J.: "Corporate Learning and Quality Control at the Bell System, 1877–1929", in: Business History Review 79 (2005), p. 39–72, here p. 39. Although Miranti focuses on Bell's US operations, similar processes and systems were implemented by Bell Canada. Miranti does not address the roles played by repair and maintenance in the development of systemic quality assurance, but their importance can be easily drawn out from his account. See ibid.

11 Agreement between the Northern Electric and Manufacturing Company Limited and the Bell Telephone Company of Canada Limited, 10 Sep. 1912, BCA, B-21121-S contract 1912.

As the number of telephones in service grew over the first decades of the 20th century, the need for workers to repair and maintain the telephone plant grew, too. While the 1912 agreement had identified telephone repair as an integral part of Bell's operations, the actual work of repair remained an ad hoc practice for some years, as Bell struggled to meet demand and maintain service quality. Northern Electric's manufacturing activities were consolidated in a single building in Montréal in 1912, but Bell continued to operate local storerooms and repair workshops (including in Toronto and Montréal) until 1925 and 1929, when Distribution Houses were built in both cities.

EARLY MAINTENANCE AND REPAIR PRACTICES AT BELL CANADA

While there are few records of Bell's repair operations prior to 1925, a series of interviews conducted in 1968 with retired Bell repairmen have proven to be invaluable sources of information about that under-documented period (and beyond). Employed over the period between 1907 and 1966 in Bell's Toronto Plant Department, they offer insights into the day-to-day routines of repairmen and how the role and practices of repair changed as the company grew.

The interviewees were all hired between the years 1907 and 1925, a period when high demand for telephone service coincided with wartime labour shortages. They describe Bell's hiring practices as being relatively informal – each of them recalled being recruited by neighbours and relatives who were Bell employees, and it seems that Bell's workshop supervisors had the authority to hire workers directly. Each of the men was hired without previous experience doing telephone work, which was not unusual given the relative newness of the industry. A job at Bell was seen as offering good, steady, dependable work. George Dumbleton, who appears to have been in his very early teens when he was hired in 1907, spent the first 18 months on the job "cleaning sets", "repairing receivers" and "doing odd jobs around the Storeroom" before he was transferred to work in installations, and then moved on to work in repairs.[12] His long sojourn in the storeroom was likely due to his young age. Harald Judson, who was in his early

---

12 George L. Dumbleton. Interviewed by A. Barwell. Life Story. Toronto, 8 Dec. 1967. BCA, Bio-file: G. L. Dumbleton. Dumbleton talks about preparing for the Owen Sound cut-over, which took place In 1959, and then goes on to describe working for a number of years after that, so he likely retired in c. 1960. If he was 14 years old in 1907, and he worked until 1960, he would have retired at 68 years of age.

20s when he joined Bell in 1922, also started as a storekeeper (at $0.35 an hour) but only stayed there for six months before being moved into installation work.[13] "Ted" Eames was hired in his mid-20s and was assigned to installations from the start, but only worked there for six months before being transferred to repair work.[14] According to Judson, installation work was "a young man's game, lots of speed, and go, go, go" and moving to repair work was seen as a step up, but the boundaries between installation and repair during these years were fairly permeable. While Dumbleton and Eames remained working in repairs once they were transferred, Judson described being reassigned to installation work at various times in his career when the need arose.

The repair shops operated on an apprenticeship model: new employees received their training in telephone repair and maintenance on the job from more experienced workers. If installation work was a "young man's game", repair work was the purview of a more diverse group of workers. Judson recalled that one of the three senior men who worked exclusively repairing and testing subscriber equipment was a war amputee, one of a number of veterans who found work at Bell after the First World War ended. Another, a part-time employee, was "an odd type" but "a magnificent pianist", who Judson believed was hired so that he could play in the Bell Telephone Orchestra.

Like the telephone industry itself, Bell's on-the-ground operations of installation and repair appear to have been somewhat informally organised during the first decades of the 20th century. Bell's Toronto repair workshop was housed in its Elm Street Storehouse, a former garage located in the downtown district. It is possible but unlikely that Bell owned the property; it appears that during this period Bell leased both the buildings and the services it used on an as-needed basis. Dumbleton, who worked at this site, reported that materials used by the repair shop were delivered to the Elm Street site by a "horse and waggon" leased from "Mr. Walsh on Queen St.". Installation crews and their equipment were also transported this way. Bell repairmen, on the other hand, were left to arrange their own means of getting to and from their job sites. The Church Street Storeroom where Judson worked was located in an old church that had been converted into a boxing gym. Bell rented part of the church for its storehouse and it seems that

---

13 Harald Dowson Judson. Interviewed by F. F. Pendock. Life Story. Toronto, 17 Sep. 1968. BCA, Bio-file: Harald Dowson Judson.

14 Edward (Ted) J. Eames. Interviewed by A. Barwell. Life Story. Toronto, 7 Feb. 1968. BCA, Bio-file: Edward (Ted) J. Eames.

much of the district's subscriber equipment – telephone sets and some PBXs – was repaired here.[15]

During these years, repair work took place at the repair workshops, at the homes of residential subscribers, and on site at the locations of Bell's commercial subscribers. Repair work that took place in the workshops typically involved repairing and reconditioning telephone sets and smaller PBXs that Bell retrieved when subscribers cancelled their service or moved. Judson explained that repairs to "the simple sets of that era" – 1293 wall sets, D020 desk stands, 295A bell boxes and 1294 wall sets – were straightforward: "[w]e changed hook switches, receivers, transmitters, cords, etc." Once repaired or reconditioned, these telephones could then be picked up by the installation men to be connected at a new location. Even with repaired telephones augmenting Bell's stock, Judson recalled that during this period the storehouses "were [often] very short of sets and installers would come in waiting for them". Judson also recalls that the shortage of telephones was often matched by a shortage of installers. He remembered that sometimes "sets" delivered to homes and offices "might be there for a couple of months before an installer came around to connect them up. Sometimes, by the time the installer got there the people who ordered the phone, had gone".[16] The shortage of workers – and the growing demand for telephone service – was likely the reason that some plant employees worked in both repairs and installations over their careers.

Bell repairmen were also sent to the homes and offices of subscribers to fix telephones that had malfunctioned. While Bell leased trucks when it needed to move materials or deliver equipment, repair workers were responsible for getting themselves and their tools to and from their work sites. Eames recalled that repairmen working in Toronto "either walked to the job or rode a bicycle. The Company paid us $5.00 a month for the use of our own bicycle. You would put your tools on the back, your wire etc. on the handlebars and either rode or pushed it to the job."[17] This was manageable if they were assigned to work in their home district; otherwise repair workers used public transit to get to the job site. Toronto was informally divided into West, Central and East districts – but

---

15 There is some discrepancy between the accounts of Dumbleton and Judson as to whether the Elm Street Storehouse or the Church Street Storehouse was Bell's first repair shop in Toronto, and whether one replaced the other or both operated simultaneously. The two sites are located no more than 1.5 km apart, approximately a 15–20 minute walk.

16 All quotes Judson, p. 2.

17 Eames, p. 4.

Bell's installation and repair operations were not organised by district at this time. Job orders were sent out from a centralised dispatch office in the downtown district and the repairmen could be sent anywhere in the Toronto territory. Judson, who worked primarily on residential installations and repairs, recalled that this sometimes resulted in workers spending much of their time commuting, not always productively, from job site to job site.

"If you were out in the West and called in, you might be dispatched to the East End and have to pedal all the way across Toronto.... When you got down there the people might be out [and you might then be] sent up [to the north end of town]."[18]

Repair workers tried to make any necessary repairs they could at the subscriber's location, both in order to save a return trip and because of the scarce supply of replacement sets. "If there was anything wrong with the telephone and it had to be changed, it was quite a job to get a new one. What we used to do was come into the garage at night, look over the telephones recovered by the installation that day and if they looked to be good, those [were] the ones that went into the subscriber's premises the next day."[19] Telephones returned from one subscriber's home or office were quickly pressed into service for the next, as demand outpaced production capacity.

According to Eames, the technologically simple telephone instruments of the period were relatively easy to repair.

"[Most] had an old 323 transmitter that used to get packed every once in a while. So we would undo the screws, loosen the button, give it a twirl with a thumb and put it back together again, perhaps blow into it a couple of times, and most of the time it would make it work all right."[20]

His account suggests that repair work often relied as much on creative improvisation as on technical skill.

"The diaphragm [of 122 or 144 type receivers] used to get rusty and you could fix that by taking it out, dusting it on the seat of your pants and then putting it back into the receiver. You would tell the subscriber that it was all right now, she would call a friend and then

---

18 Judson, p. 3.
19 Eames, p. 4–5.
20 Ibid., p. 6.

usually tell you that it was now just 'wonderful'. That would be the extent of many repairs on telephones."[21]

While all three repairmen downplayed the degree of technical skill required to repair residential subscriber equipment, each of their accounts reflects an awareness of the importance of maintaining good relations with subscribers.

Eames described residential repairs as typically routine, but recalled having a bad experience when a subscriber falsely accused him of taking her purse and "[didn't do] much apologizing" when she later found it.[22] Eames found residential repairs neither interesting nor challenging. He was happy when he was sent to work with Bell's business subscribers in Toronto's downtown district, where he was able to maintain and repair a wider range of telephone equipment, including switchboards. Judson, on the other hand, enjoyed the variety and sociability of residential repair work, noting that "there [was] something different all the [time] as no two jobs [were] alike". He observed that it was the nature of repair work that repairmen had to deal with subscribers who were not at their best. "Our reception [by the subscriber] was not always good with the phone out of order. However on installation, if they [had] been waiting a long time for a phone, your reception [was] no better. There is a great sameness about installation and a great variety on repair."[23] Apparently, the varied jobs Bell's residential installation and repair workers were called on to do included rescuing cats that climbed up telephone poles and couldn't find a way down. Eames described the established protocol for accomplishing this job without sustaining injuries as wearing a heavy coat, making sure to turn up the collar, climbing up the pole and waiting for the cat to jump on one's shoulder, and then climbing back down, at which point the cat would typically make its happy escape. He claimed to have successfully rescued many cats this way without receiving a scratch.[24]

Senior and more skilled workers were typically assigned to the repair and maintenance work done on site for Bell's larger commercial subscribers, such as

---

21 Ibid. In his ethnographic study of the field service work performed by Xerox technicians, Julian Orr notes that technicians considered the maintenance of good relations with their customers as being just as important as the maintenance of the machines. He describes how maintaining this "triangular relationship of service" between technicians, customers and machines allowed technicians to establish an image of authority and professionalism. Orr, Julian: Talking about Machines: An Ethnography of a Modern Job, Ithaca: ILR/Cornell University Press 1996, p. 78–79.

22 Eames, p. 5.

23 Both quotes Judson, p. 5.

24 Eames, p. 6–7.

hotels, department stores, and railway stations. This work could include repairing any malfunctioning telephone sets, but primarily involved keeping the switchboards of commercial subscribers in good working order. Maintaining PBXs required a greater degree of technical expertise and these assignments were viewed as a sign of an employee's greater reliability and superior skills, and therefore higher status. Dumbleton explained that when he "graduated" from repairing sets, he spent several years assigned to the maintenance of switchboards for both Canadian Pacific Railway and Canadian National Railways at Toronto's Union Station, and then did similar work at the King Edward Hotel.[25] Two of the three interviewees described having worked at Eaton's, Canada largest department store, installing and maintaining the switchboards and "order tables" that were required for its business and telephone sales operations (fig. 2).[26] Eames' description of the repair work he did in the downtown district in the 1920s demonstrates both the rudimentary nature of the era's telephone technology and the degree to which ongoing maintenance and repair were critically important to its proper functioning. Many downtown businesses and offices were fitted with 550- and 551-type switchboards, "most of which were battery [fed] from central offices on cable pairs with varying loads. The voltages used to go up and down like a yo-yo. Subscribers had much trouble with this and to overcome this the first thing [Bell] did was to send out a reconditioning crew to current flow all the relays." Apparently, the problems with these switchboards were not easily resolved: Eames noted that "[s]ome of the men did that [work] for months and months".[27] Eames estimated that, in order to ensure a reliable telephone service for its business subscribers, Bell had approximately 60 men working in repair and maintenance at this time in the Toronto downtown area alone.

By virtue of the fact that Bell's commercial subscribers required regular service and maintenance, repairmen sometimes found themselves playing a policing role for Bell. Dumbleton described discovering "buckshee" (or "free") connections on his service calls.

"One day ... I was repairing one phone and from my position could see another three or four phones around me. I heard someone talking but could not see what phone was being used or who was using it. So I took a good look around and found that one of our ex-employees had installed a Stromberg-Carlson switchboard for them and placed phones on every floor, having run a concealed wire to one of our trunks. When I disconnected their

---

25 Dumbleton, p. 1.

26 Eames, p. 9; Judson, p. 1.

27 Both quotes Eames, p. 8.

concealed wiring, they put up quite a fuss. Later when I checked it again I found it had been reconnected so this time I reported it ... and after some fussing around with various people we got straightened away."[28]

Figure 2. Operators working at the "order tables" at the Eaton's department store. Bell repair and maintenance workers kept switchboards such as these in good working order (Montréal, 1931, BCA, A-30147).

It is noteworthy that Dumbleton chose not to report the unauthorised connections in the first instance. Whether this was due to a reluctance to inform on a former colleague or a gesture of good faith in the business subscriber's integrity, Dumbleton clearly believed he possessed a significant degree of autonomy in his role as a repairman.

While most of Bell's commercial subscribers were businesses located in urban centres, some were located well beyond the limits of the city's public transit network and further than one could reasonably expect a repair worker to travel by bicycle. Dumbleton remembered that when he and his co-workers were sent

---

28 Dumbleton, p. 4.

out to make repairs at Harris' Slaughterhouse, which was located about three-and-a-half kilometres east of the city, "we had to take the streetcar [most of the way there], and then Harris would send someone with a pony and gig to take us the rest of the way".[29] When they were dispatched to do repair work at Taylor's Farm, located to the northeast of Toronto, "we had to take a train so far and then hike it the rest of the way".[30] Judson described the lengthy journey he had to make when he was sent up to do repairs at the Langstaff Jail Farm, approximately 25 km north of the city: he had to pedal up to the north edge of the city, "leave my bicycle there, and take the Radial Car [suburban railway] from there. One of the boys at the prison farm would come out to the Radial to pick [me] up and take [me] back".[31] Bell's apparent lack of concern about the inefficient use of its workers' time underscores the importance it placed on keeping its service in good working order and its subscribers happy. It speaks, too, of Bell's expanding network and the growing ubiquity of telephone service among Canada's urban and suburban populations.

RATIONALISING TELEPHONE MAINTENANCE AND REPAIR

As Bell reorganised its operations according to the principles of Scientific Management, new cadres of professional managers, often engineers, began to oversee and then replace the foremen who had been responsible for running the storerooms and workshops.[32] Repair work moved from Bell and Northern Electric's workshops to new modern purpose-built facilities. The new Toronto Telephone Distribution House was erected in October 1925, and its Repair Department opened in the autumn of 1926.[33] Montréal's Telephone Distribution House was completed in early 1929, with its Repair Department installed in February of that same year.[34] While the apprentice model of on-the-job training for novice telephone repair workers continued, it was shaped and supplemented by courses

---

29 Dumbleton, p. 5.
30 Ibid., p. 5.
31 Judson, p. 4.
32 Miranti, "Corporate Learning", p.46 and 48.
33 Telephone Distributing House Premises–Toronto (1925), BCA, Northern Electric Historical Collection, Vol. 4, #10791.
34 Telephone Distributing House Premises–Montréal (1929), BCA, Northern Electric Historical Collection, Vol. 4, #10508.

that were mandatory for both new and experienced workers.[35] Dumbleton recalled taking and enjoying "several of the courses", as did Judson.[36] Eames described the implementation of weekly meetings at this time. Led by the Repair Department manager and foreman, the meetings were designed to keep employees informed of technical advances and changes and to convey job safety guidelines and procedures.[37]

Another effect of Bell's adoption of Scientific Management methods was a sharp increase in the production and circulation of information about Bell's and Northern Electric's operations. In addition to thorough documentation on operational activities and finances in their internal communications, both companies also introduced publications aimed at keeping employees – and, in the case of Bell, its subscribers as well – informed about new developments.[38] Whereas it was difficult to find information about Bell's repair and maintenance practices prior to the mid-1920s, by the 1930s, feature articles in these new publications offered detailed accounts and photo documentation on repair operations in the Toronto and Montréal Distribution Houses (fig. 3).

According to a 1938 article in Bell Canada's *Blue Bell* magazine, each Distribution House contained a warehouse, a repair shop, and a company garage that housed a fleet of 100-plus trucks used to pick up returned materials.[39] It portrays repair operations at the Distribution Houses as fully routinised, with repair shop workrooms that were modelled on laboratories: orderly, clean and well lit, with fully-equipped work benches. It details the journey that "returned" telephones took from their arrival at the Distribution House through the assessment process that determine whether they would be "junked" or sent to the Repair Shop. Returned telephones deemed worthy of repair were sorted again in the Repair Shop

35 Dumbleton was one of the repairmen who were often assigned to train new employees. In his interview, he recalled that during this period there were "two years steady [when] I don't think there was a week without having someone with me to train. I finally got fed up with the whole thing". Dumbleton, p. 4.
36 Ibid.; Judson, p. 5.
37 Eames, p. 8–9.
38 For example, Bell began publishing Blue Bell magazine in 1921 to keep its employees informed about new developments and company social events. Northern Electric began publishing its employee magazine, Northern Electric News, in 1927. Bell also began publication of a circular called Telephone News in 1934 to inform subscribers of new developments and proper telephone etiquette.
39 Telephone Repair Shops, Blue Bell, (Jun. 1938), p. 16–18.

according to the types of repair needed, before being repaired, tested, then packaged and sent to the warehouse to await future installation.[40]

Figure 3. "Alvin Keith checks a consignment of miscellaneous material and equipment sent from the Bell Telephone distributing room to the returned material department or the Northern Electric Company on Shaw Street." (Image and caption appeared in Blue Bell, Bell Canada's employee magazine, Oct. 1942, BCA, A-38716-01)

The 1938 article is significant in that it provides a careful account of the volume and value of Bell's repair operations – information that could be inferred from earlier documents but not confirmed. The article reported that the telephone apparatus returned to the Toronto and Montréal repair shops in 1937 had included: 173,000 telephones, 74,000 bell boxes, 460 PBX switchboards and 670 telephone booths, with an estimated total value of C$4,676,000. Of the total number of returned apparatus, 152,000 telephones, 62,000 bell boxes, 304 PBX switch boards, and 548 booths (or 65 to 85% of the devices) were repaired and/or reconditioned, and moved into storage at the Telephone Contract Division's warehouses, ready to be put back into service. According to the article, the recuperat-

---

40 Telephone Repair Shops, p. 16.

ed value of the repaired apparatus was C$3,834,320 at a cost of C$748,160 for parts and labour.[41] To put this into perspective, Bell's repair operations yielded more than C$3,000,000 of recuperated value in a year when Bell's reported operating balance was C$12,519,975.55.[42]

Figure 4. "These [wall sets] have been repaired and reconditioned and are being cleaned before being refinished. The Cleaners are left to right, John Wilkinson, Gordon Hann and James Dunn." (Image and caption appeared in Blue Bell, Bell Canada's employee magazine, Oct.1942, BCA, A-38716-04)

The representation of the value of repair and maintenance in this article – and in similar articles and reports that followed in the 1940s and 1950s – is both a demonstration and an effect of Scientific Management's influence on the operations of Bell and Northern Electric's Telephone Contract Division beginning in the 1920s. While these later records offer detailed descriptions of the *fiscal* value of repair and maintenance to Bell's operations, evidence of Bell's recognition of the *instrumental* value can be inferred from its integration of repair and mainte-

41 Telephone Repair Shops, p. 18.
42 Bell Telephone Company of Canada Annual Report, 1937.

nance in its earliest conception of its telephone business. By recuperating, repairing and reconditioning, rather than replacing, its telephone sets and equipment, Bell's repair operations proved to be an important factor in the company's ability to manage costs, meet demand for service, and monitor and maintain acceptable levels of technological quality (fig. 4).

CONCLUSION

The accounts of the Bell Canada employees offer insights into the day-to-day routines and experiences of telephone repair work in Canada's largest cities during a period for which little documentation exists. Repair work was perceived by these employees as being of higher status than installing telephones, requiring technical knowledge and resourcefulness as well as social skills. They recognised and seemed to enjoy the fact that they possessed a degree of independence that was greater than other workers in the Plant Department. Taken together, the recollections of the repairmen show repair work to have been a necessary and integral component of the telephone as a technology-in-use, which literally ensured that telephones and the telephone network could be "used". They describe how Bell's early telephone repair and maintenance operations not only ensured that the telephone system functioned properly; they also acted as a stopgap that allowed the company to both meet subscriber demand and bridge intermittent shortages of equipment and workers. Just as importantly, they also describe how the practices and role of repair changed over the course of their careers, with the introduction in the 1920s of more systemic organisation and management across all Bell System operations, including those of its Canadian branches, Bell Canada and Northern Electric.[43]

While there is little evidence that environmental concerns motivated Bell's early decision to lease and repair its telephone equipment, it is interesting to consider what the ecological effects of that decision might be if it was still in effect today, a little more than one hundred years later. What if today's (mobile) telephones – purchased, used and disposed of, like any other commodity – were designed to be returned, repaired and reconditioned for reuse? The reflection of a Northern Electric employee, interviewed in 1983, offers an example of the persistence of technology in the *post-liberalised* telephone industry, and another reason for historians of technology to shift their focus from innovation to repair. He observed that if

---

43 Miranti, "Corporate Learning", p. 47.

"Northern ceased to offer support for its old equipment it would mean that every time an old piece of equipment would [break] down, we would have to throw it out rather than repair it. [They used to] say the product life is 40 years [but] it's a lot more than that, because we're still maintaining switchboards that were installed long before 1925 and they're still working and they're still giving good service".[44]

This employee's observations suggest that maintenance and repair may in fact be more durable ideas – perhaps even more modern, and more profitable – than planned obsolescence. This chapter suggests that looking back at the telephone industry's practices of recuperating, repairing and reconditioning, rather than replacing its telephone sets and equipment, may offer a way to imagine the role of maintenance and repair in the lives of our present-day technological devices.

---

44 Jack Brighton. Interviewed by Gordon Bennett. Life Story. Jan. 1983. BCA, Bio-file: Jack Brighton, p. 6.

USERS AND REPAIR

Mud Bricks in a Concrete State:

Building, Maintaining and Improving One's Own House
in Soviet Samarkand, 1957–1991

Jonas van der Straeten and Mariya Petrova

In 1963, the journal *Building and Architecture in Uzbekistan* printed an arranged photo of four urban planners engaged in an animated conversation about an architectural model in front of them.[1] This model displayed the prospective city centre of Samarkand after its socialist transformation. In the model, the city's old Islamic neighbourhoods had been entirely demolished and replaced by multi-storey, prefabricated residential buildings that lined roads radiating from the historical Registan complex. These neighbourhoods of the Old City, known as *mahallas*, had long been a thorn in the side of Soviet planners – symbolically and materially the narrow alleyways and mud-brick houses stood in the way of the city's modernisation. Now, after having barely changed in appearance for several hundred years – throughout half a century of Russian colonial administration and thirty years of Soviet Stalinist rule – the *mahallas* were set to finally give way to the rational urban planning of the Khrushchev era. This model of the new city centre, however, never became reality, owing to resource constraints and controversies over the final concept for its reconstruction.[2] Ultimately, the *mahallas* would even survive the mass housing campaign that started in the late 1950s. The overall vision of an all-out modernisation of Samarkand remained confined to a handful of micro-districts (*mikrorayons*) on the outskirts of the city. Until the collapse of the Soviet Union and beyond, Samarkand would large-

---

1   Printed in Stroitel'stvo i arhitektura Uzbekistana, 01 (1964), p. 34.
2   Central State Archive of the Republic of Uzbekistan/ Tsentral'nyj Gosudarstvennyj arkhiv Respubliki Uzbekistan (TsGARUz), f. 2532 (Architects' Union of the Uzbek SSR/Sojuz Arkhitektorov UzSSR), op. 2, d. 21, ll. 61–72.

ly remain a city of adobe bricks, with prefabricated concrete apartment blocks at its fringes only.

By and large, research on Soviet cities has focused on change and the transformative power of socialist urban development.[3] By looking at the Soviet modernisation project in Central Asia through the prism of Samarkand, a second-tier city in the Uzbek SSR, our chapter draws attention to a degree of (ethno)cultural and material persistence in the cityscape that seems surprisingly high, given the Soviet Union's own aspirations regarding urban modernisation. Cultural and material persistence, we argue, cannot be regarded independently of each other. On the one hand, the courtyard house as a type of urban housing that is widespread in the cities of Central Asian oases embodies local (ethno)cultural norms and conventions associated with the social production of space.[4] Adobe brick architecture proved to be well suited to address these requirements for space-making and hence persisted as the dominant technology for the maintenance and renovation of family houses, especially in light of a general scarcity of industrially manufactured construction materials. On the other hand, the inherent characteristics of this millennia-old technology, its intensiveness in terms of labour and maintenance, demanded recurrent collective activities of rebuilding and renovating. The culture of collective self-help that was a precondition for "individual house construction" (*Individual'noye stroitel'stvo*) did not only serve a functional purpose.[5] For some local ethnic groups it can be interpreted as an important

---

3 See Kotkin, Stephen: Magnetic Mountain. Stalinism as a Civilization, Berkeley: University of California Press 1997; Stronski, Paul: Tashkent. Forging a Soviet City, 1930–1966, Pittsburgh: University of Pittsburgh Press 2010.

4 Ethnic designations are a particularly intricate issue, yet they remain an important heuristic. In our text, for example when referring to Tajiks, Uzbeks and Bukharian Jews as dwellers of the Old City in Samarkand, we use designations that correspond to the (context-specific) self-attribution of our respondents (or their voices in the archives). When doing so, we are well aware of the historical complexity of the linguistic and ethnic landscape in Central Asia in general and in Samarkand in particular, as well as the fact that existing ethnic categories were defined and enforced by the Soviet administration as part of the national delimitation politics. For an in-depth discussion see: Hirsh, Francine: Empire of Nations. Ethnographic Knowledge and the Making of the Soviet Union, Ithaca: Cornell University Press 2005; Suny, Ronald Grigor/Martin, Terry (eds.): A State of Nations. Empire and Nation-Making in the Age of Lenin and Stalin, New York: Oxford University Press 2001.

5 "Individual house construction" is an official Soviet term, based on the legal division between "private" and "individual" property, whereby the former was a "capitalist" form of property, used for generating private profit, and the latter was conceptualised

act of cultural self-affirmation – not least through the reference to *hashar*, a tradition of neighbourhood help deeply engrained in narratives about Central Asian traditional heritage and cultural identity.

Following Stephen Kotkin's call to "[shift] the focus from what the party and its program *prevented* to what they made *possible,* intentionally and unintentionally", we therefore offer a perspective on the fragmentary nature of the Soviet urban modernisation process in Central Asia that focuses on the frameworks of individual action.[6] The persistence of privately built adobe brick houses in Samarkand, we argue, does not simply mirror the deficiencies of the Soviet shortage economy and the limitations of Soviet urban development. By applying elaborate strategies of self-help and labour mobilisation in construction, the residents of Samarkand maintained considerable agency not only in shaping their material environments but also in preserving "traditional" (and hence pre-Soviet) cultural identities.

Our analysis draws on in-depth oral history interviews with owners of houses and apartments in a variety of neighbourhoods and their corresponding modes of housing in Samarkand.[7] These are complemented by archival research in the City Archive of Samarkand and the Central State Archive in Tashkent as well as a review of contemporary Soviet literature on construction and architecture in Central Asia, for example the journal *Building and Architecture in Uzbekistan*

---

as property for personal use only and therefore as legitimate in a socialist society. See Smith, Mark B.: Property of Communists. The Urban Housing Program from Stalin to Khrushchev, DeKalb: Northern Illinois University Press 2010, p. 143. When referring to the practice of building one's own house (usually a free-standing one-family house) for personal use – especially in the context of the Soviet legal system –, we therefore use the term *individual house building* in our chapter. For the sake of readability, we replace this expression with more commonly used terms like "privately owned houses" and "self-help construction" when the legal context is not of relevance.

6 Kotkin, Magnetic Mountain, p. 22 (italics original).

7 For our study, we conducted a total of ten in-depth oral history interviews with residents living in different neighbourhoods of Samarkand. Our qualitative approach does not, of course, provide for statistical representativeness and extrapolation to all strata of urban society. Yet while being aware of the pitfalls of oral history, such as skewed and incomplete memory, we were surprised by the congruousness and consistency of responses regarding self-help building and its cultural interpretations. Our educated guess is that mundane practices of house building leave traces in memory that are less blurred by shared narratives than, say, the impacts of political events. Moreover, most of our interviews took place within the very houses that were the subject of interest, and through their very materiality they served as a handy memory aid.

(*Stroitel'stvo i arkhitektura Uzbekistana*).[8] The chapter focuses on the last three decades of Soviet rule in Uzbekistan, between 1957, the year that marked the shift from a rather reactive housing policy to a proactive policy under Khrushchev, and the year 1991, the end of the Soviet Union and its economic and legal framework for housing construction.

BUILDING AND IMPROVING HOMES IN LATE SOVIET SAMARKAND: FRAMEWORKS OF INTERPRETATION

Our micro-study on the persistence of adobe brick houses in Samarkand and the people building, maintaining and improving these houses connects with a debate that revolves around the question of continuity and persistence in the process of Central Asia's Sovietisation. In historical research on technology and material culture, a field that has long been preoccupied with change and novelty, continuity and persistence are two analytical categories scholars have barely made use of.[9] Research on Central Asia is no exception. The little attention the region has received regarding its history of technology has been almost exclusively focused on the transformation of societies and environments brought about by the transfer of large-scale technologies under Tsarist Russia, and later under Soviet rule. Matthew Payne has described the Turksib railroad in terms of a grand social engineering project, the "forge of the Kazakh proletariat" to transform nomads into an industrial working class.[10] In the tradition of James Scott, an extensive body of literature has investigated the large-scale irrigation schemes that turned the

---

8   Uzbek archives, although difficult to access for foreign researchers, are generally well organised and hold a wealth of information left by a Soviet administration almost obsessed with written documentation. The voices of private house owners and house builders can be traced especially well through documentation left by the lowest level of state administration (e.g. neighbourhood committees) or the widespread practice of writing petition letters.

9   Edgerton, David: The Shock of the Old. Technology and Global History since 1900, London: Profile Books 2006. For an overview see Krebs, Stefan/Schabacher, Gabriele/Weber, Heike (eds.): Kulturen des Reparierens: Dinge – Wissen – Praktiken, Bielefeld: transcript 2018.

10 Payne, Matthew J.: Stalin's Railroad. Turksib and the Building of Socialism, Pittsburgh: University of Pittsburgh Press 2001.

Central Asian steppe into one of the biggest cotton-growing areas of the world – and examined these schemes' disastrous environmental (and social) impacts.[11]

According to the narrative underlying most studies, Soviet rule in Central Asia profoundly reconfigured every aspect of the material foundation of state, sociality and everyday life.[12] In this narrative, Central Asians only appear as actors once they have been incorporated into the Soviet state economy as workers, technicians and later also engineers or specialists – especially in the two post-World War II decades, when the first generation of Central Asians had passed through the Soviet higher education system. What these studies largely ignore, however, is the persistence of technologies that are termed "local" or "traditional" and their interaction with the Soviet project. The mundane realm of local house building, we argue, is fertile ground to which the analytical lens of persistence and continuity in the Sovietisation of Central Asia can be applied.

While still prevalent in the field of technology, the interpretation of Soviet rule as a top-down modernisation process has been questioned and differentiated in historical research on topics such as religion, gender or the formation of a Central Asian Soviet intelligentsia.[13] While the Stalinist period has been extensively studied and arguably considerably shaped the image of Soviet Central Asia, the period after Khrushchev's accession to power, when the Soviet Union showed its more humane face, has attracted much less attention. A laudable ex-

---

11 See for example Obertreis, Julia: Imperial Desert Dreams: Cotton Growing and Irrigation in Central Asia, 1860–1991, Göttingen: V&R Unipress 2017.

12 Van der Straeten, Jonas: "Borderlands of Modernity. Explorations into the History of Technology in Central Asia, 1850–2000", in: Technology and Culture 60, 3 (2019), p. 659–687.

13 Roberts, Flora: Old Elites under Communism. Soviet Rule in Leninobod, PhD thesis, University of Chicago 2016, p. 19; see also Khalid, Adeeb: Making Uzbekistan. Nation, Empire, and Revolution in the Early USSR, Ithaca: Cornell University Press 2015; Khalid, Adeeb: Islam after Communism. Religion and Politics in Central Asia, Berkeley: University of California Press 2007; Northrop, Douglas: Veiled Empire. Gender and Power in Stalinist Central Asia, Ithaca: Cornell University Press 2004; Kamp, Marianne: The New Woman in Uzbekistan. Islam, Modernity, and Unveiling Under Communism, Seattle: University of Washington Press 2006; Kalinovsky, Artemy: Laboratory of Socialist Development. Cold War Politics and Decolonization in Soviet Tajikistan. Ithaca: Cornell University Press 2018; Stronski, Tashkent; Florin, Moritz: Kirgistan und die sowjetische Moderne: 1941–1991, Göttingen: V&R Unipress 2015. Florin analyses public and elite discourses in Kyrgyzstan, showing the inner-republic perspective on important historical, political and cultural events and processes in Soviet post-war history.

ception is Sergey Abashin's book *Soviet Kishlak*, a historical ethnography of one village in northern Tajikistan, which is highly instructive for getting beyond the established dichotomies of "tradition" versus "modernity", "resistance" versus "accommodation" – not least because the author is one of the few to transcend the boundary between Soviet and Western scholarship.[14] Abashin pays particular attention to the complexity of the often contradictory relationship between the Soviet state and its citizens. While certain Soviet practices became part of the daily life of villagers, he shows, they coexisted and overlapped with other practices that were considered "Muslim", "national" or "traditional". Most *kolkhoz* members, for example, worked on collective lands only occasionally and dedicated most of their time to other activities such as animal husbandry or handicrafts. We follow Abashin's suggestion to conceptualise local people's living environment as a "mosaic, consisting of multiple sub-spaces", not only for rural Tajikistan but also for Samarkand.[15] While Abashin is primarily interested in social spaces, we look at the material implications of this fragmentation and the agency of individual people in shaping materiality within different sub-spaces.[16]

Housing appears as a specific field of study in this regard. On the one hand, it was an area of state intervention that saw the most ambitious (and arguably successful) political programmes, in particular the mass housing programme under Khrushchev starting in the late 1950s, and the most far-reaching social engineering visions of changing people's daily lives and material conditions. On the other hand, it was one of the policy areas in which the state administration was arguably most reflective about its limitations in terms of resources and institutional efficiency and allowed for a relatively high degree of permitted autonomy and *individual* ownership.[17] It is no irony that Stephen Kotkin's widely cited call to study the agency of individuals within Soviet society is taken from a book on what was arguably one the Soviet Union's most radical urban development projects.[18] In contrast to Kotkin, who studies an urban and industrial structure that was built from scratch, our chapter looks at those frameworks of individual action that allowed for the persistence of traditional cultural practices and materials in urban environments in spite of the Soviet modernisation project.

---

14 Kalinovsky, Artemy: "Exploring 'Sovietness' in Local Context", in: Central Asian Affairs 4 (2017), p. 293–304; The essay is a book discussion of Abashin, Sergei: Sovetskii kishlak: mezhdu kolonializmom i modernizatsiei, Moscow: Novoe Literaturnoe Obozrenie 2015, p. 293–296.

15 Abashin, Sovetskii kishlak, p. 610.

16 Ibid.

17 See also Smith, Property of Communists.

18 Kotkin, Magnetic Mountain, p. 22.

Our emphasis on persistence shows the limitations of a term that is common-ly applied to activities of self-help home improvement in modern industrial soci-eties: "Do-it-yourself" as a "culture" or even "movement" has been widely de-scribed as an inherently novel phenomenon, exported from the US to Western Europe in the 1950s and intimately tied to (and made possible by) the emergence of the affluent post-war society and the specific consumption culture and (male) leisure culture associated with it.[19] In research on the Soviet Union, too, authors have established a qualitative change in the Soviet "repair society" following the "consumer turn" initiated by Khrushchev in the 1950s.[20] The latter gave rise to an amateur "do-it-yourself" culture in the 1960s that has been interpreted in dif-ferent ways – either as a partial subversion of Soviet state ideology through its emphasis on individualism[21] or, in contrast, as firmly anchored in both the insti-tutional and symbolic universe of the Soviet Union.[22] When applied to the pe-riphery of Soviet Union, however, the idea of a "do-it-yourself" culture arguably loses much of its explanatory power. The practices of building, maintaining and improving private homes in Soviet Samarkand that we describe in this chapter are not discursively framed as "modern" or "Soviet", nor are they manifestations of an individualist leisure culture within a collectivist system. On the contrary, they stand out for the references made in their interpretation to a *pre-Soviet* tra-dition of collective self-help.

SOVIETNESS AS A MOSAIC: THE TRANSFORMATION OF SAMARKAND'S CITYSCAPE UNTIL 1960

Our study focuses on the traditional courtyard house, the most prevalent type of housing in the cityscape of late-Soviet Samarkand. This prevalence was first a

---

19 On Germany see Voges, Jonathan: "Selbst ist der Mann": Do-it-yourself und Heim-werken in der Bundesrepublik Deutschland, Göttingen: Wallstein 2017, p. 10–12.

20 Golubev, Alexey/Smolyak, Olga: "Making Selves Through Making Things. Soviet Do-It-Yourself Culture and Practices of Late Soviet Subjectivation", in: Cahiers du monde russe 54 (2013), p. 517–541, here p. 526. See also Zinaida Vasilyevas' (yet unpublished) thesis project on Do-it-yourself and amateur culture in late and post-Soviet Russia. For an interim report see Vasilyeva, Zinaida: "Do-It-Yourself Practices and Technical Knowledge in Late Soviet and Post-Soviet Russia", in: Tsantsa 17 (2012), p. 28–32.

21 Siegelbaum, Lewis H.: Cars for Comrades. The Life of the Soviet Automobile, Ithaca: Cornell University Press 2011, p. 243.

22 Golubev/Smolyak, "Making Selves Through Making Things", p. 521.

result of the city's pre-Soviet architectural legacies. The Tsarist administration, after its conquest of Samarkand in 1868, had demolished the citadel and burned down parts of the bazaar as punishment for acts of resistance,[23] but did not intervene in the centuries-old dense network of narrow alleyways and adobe brick courtyard houses in the old Islamic city.[24] To build a new colonial city, the imperial administration instead focused on the old citadel and the area west of it. By the time of the October Revolution in 1917 Samarkand consisted of two parts of roughly the same size: the colonial city with its geometrical layout, around a series of radial axes, starting at the old citadel. This part of the city hosted administrative and European-style houses. Interestingly, many of these houses were also built of adobe bricks, sometimes coated with burnt bricks to reinforce them.[25] The residents of these houses were merchants and families associated with the colonial administration, in total around 15,000 people of mostly European or Tartar origin. The majority of Samarkand's population of about 90,000 people, however, lived in the Old City.[26]

For the short period between 1924 and 1930 Samarkand became the capital of the newly founded Uzbek SSR, but it subsequently lost this status to Tashkent, the previous capital of the Turkestan Governorate and the Turkestan Autonomous Soviet Socialistic Republic, which existed between 1918 and 1924. The more industrialised, Russified and more Uzbek Tashkent seemed better suited for the Soviet vision of creating a socialist showcase city in Asia[27] and became the place where planning expertise for Uzbekistan and most construction resources were concentrated in the ensuing years. In Samarkand, the imprint of

---

23 Morrison, Alexander: Russian Rule in Samarkand 1868–1910. A Comparison with British India, Oxford: Oxford University Press 2008, p. 24–25.

24 According to the first all-Empire census in 1897, by that time Samarkand was mostly inhabited by Tajik-speaking Muslims and Jewish groups (36,845 people) and 5,514 speakers of Uzbek. See Nikolaj, Trojnitskij; Pervaya vseobshaya perepis' naseleniya Rossijskoj Imperii, 1897 g., LXXXIII· Samarkandskaya oblast', St. Petersburg: izdanie Tsentral'nogo statisticheskogo komiteta ministerstva vnutrennikh del 1905, available online at https://www.prlib.ru/item/436672 (accessed 14.06.2019), p. 136–137.

25 Nil'sen, Vladimir: U istokov sovremennogo gradostroitel'stva Uzbekistana XIX – nachalo XX vekov, Tashkent: Izdatel'stvo literatury i iskusstva imeni Gafura Gulyama 1988, p. 106.

26 Giese, Ernst: "Transformation of Islamic Cities in Soviet Middle Asia into Socialist Cities", in: French, Richard A./Hamilton, F.E. Ian (eds.): The Socialist City. Spatial Structure and Urban Policy, Chichester et. al.: Wiley 1979, p. 145–166, here p. 151.

27 Stronski, Tashkent, p. 18.

Stalinist rule on the cityscape remained modest except for the university and a number of dispersed residential and public buildings in the colonial part.[28] The ambitious first general plan for Samarkand from 1937-38 that provided for the Old City to be demolished and replaced by a unified, planned city centre could not be realised owing to multiple constraints, meaning that the building stock of Samarkand remained almost unchanged until the early post-war years.[29]

With the outbreak of the Second World War, any efforts for the planned transformation of Samarkand's cities ground to a halt. All of a sudden, the city had to accommodate not only several factories that had been relocated from the frontlines in Europe, but also the biggest portion of the 160,000 evacuees who had been allocated to the Samarkand region (oblast) and thousands of deportees from all over the Soviet Union.[30] In the absence of building materials, almost all of which were channelled into the war economy, the only way of accommodating the newcomers was the densification of existing housing stock. House owners all over the city were required to provide rooms to evacuees, especially in the colonial city backyards, which filled up with rudimentary structures, mostly simple shacks to accommodate whole families. These common yards (*obšiy dvor*), where facilities like water taps and toilets were shared, were among the most visible manifestations of a massive housing shortage that lasted well into the 1950s, with an average of just 5.6 m² per capita.[31] Late Stalinist strategies to alleviate the housing crisis were generally piecemeal and incoherent, while the bulk of investments were directed to industry and architects remained focused on planning original buildings of neoclassical grandeur.[32]

Between 1951 and 1957, authorities began to address the housing crisis in a more focused and systematic way than before.[33] Under Khrushchev the development of a mass housing programme gained traction and culminated in the decree of 1957, in which the leadership committed to providing separate and en-

---

28 Diener, Christa/Gangler Anette: Städte Usbekistans zwischen Tradition und Fortschritt. Städtische Transformationsprozesse der zentralasiatischen Städte Taschkent und Samarkand, Cottbus: BTU, Lehrstuhl Städtebau und Entwerfen, 2006, p. 210.

29 Ibid., p. 210.

30 Muminov, Ibragim: Istoriya Samarkanda v 2 tomakh. T. 2: Ot pobedy velikoj oktyabrskoj sotsialisticheskoj revolyutsii do nashikh dnej, Tashkent: Fan 1970, p. 232.

31 TsGARUz, f. 1619 (Central Department for Statistics/Tsentral'noe statisticheskoe upravlenie), op. 16, d. 4, l. 25.

32 Smith, Property of Communists, p. 32–42.

33 Ibid., p. 21.

closed living space for every family in the Soviet Union within a decade.[34] Building in large quantities and adopting an approach based solely on "rationality" became the key motifs of a mass building programme that was now almost exclusively based on the prefabrication of standardised multi-storey apartment buildings, commonly known as *khrushchevki*.[35] As a result of the programme, per capita housing construction in the USSR was by far the highest in Europe between 1956 and 1963. It became known for improving the lives of tens of millions of citizens and for mass producing a newly uniform and undifferentiated Soviet cityscape.[36] In Samarkand, however, it did so only at the city's Western margins. A first *Kombinat* for prefabrication of panels opened in 1958 as a precondition for a series of comparatively small housing projects to be built in the early 1960s. By the end of the 1960s Samarkand saw the completion of its first two micro-districts, ideal self-contained neighbourhoods that Soviet planners hoped would reconcile the maximisation of housing production, the need for separate family apartments and the communist idea of communal living.[37]

According to an inventory report by the city administration in 1951, 95% of dwellings in Samarkand were made of adobe bricks (70%) or timber-frame structures (with adobe filling) (25%).[38] The mass housing programme and construction of prefabricated apartment blocks from the 1960s onwards did change the proportions, but the share of houses built out of clay must have remained at least as high as the rate of privately owned houses.[39] In the republic as a whole, the rate of privately owned houses remained between 35 and 40% until the end of the Soviet Union.[40] This was not only because the old city centre remained largely unaffected by the programme, but also because the increasing volume of

---

34 Decree by the Central Committee of the Communist Party of the USSR from 03.07.1957. Available online at http://www.libussr.ru/doc_ussr/ussr_5213.htm (accessed 14.06.2019).
35 Smith, Property of Communists, p. 60.
36 Ibid., p. 4, 113.
37 State Archive of Samarkand Province/Samarkandskij Oblastnoj Gosudarstvennyj Arkhiv (SamOGA), f. 26 (Communal and Housing Commission of the City Executive Committee/Zhilishscno-kommunal'naya komissiya Gorispolkoma), op. 1, d. 2465. ll. 23–25.
38 SamOGA f.26, op. 1, d. 1363, l. 3
39 Because of prohibitive prices and restrictions on the number of burnt bricks that could be purchased in a year, clay remained the only available building material for individual houses.
40 Andrusz, Gregory D.: Housing and Urban Development in the USSR, London et. al.: Macmillan 1984, p. 291.

new private housing space after 1950 was still largely built with clay.[41] At the outskirts of the city and around the micro-districts, a thick belt of individually built adobe brick houses emerged. During the 1950s, this high degree of individual house building was not necessarily in conflict with the official Soviet housing policy. Historians have taken little note of the conspicuous policy of state-backed individual house construction that formed part of the administration's strategy to mitigate the post-war housing crisis and was a component of the mass-housing programme after 1957.[42] Builders of private houses had access to specific loans, and handbooks with instructions for the self-help construction of houses circulated all over the Soviet Union.[43]

In Samarkand, the widespread practice of individual house building even continued after 1962, when an official degree introduced much stricter rules on the issuance of plots in urban areas of the Soviet Union.[44] At the same time, the administration of Samarkand lacked effective control over building standards, regulations and the distribution of plots. Chaotic planning and disputes over jurisdiction, for example between the city administration and its neighbouring *kolkhozes*, meant that there was considerable scope for unregulated house building and extension.[45] Like their counterparts in the Old City, owners of privately built houses in the new areas for private housing and in the outskirts of Samarkand largely maintained a high degree of agency in shaping the material and spatial configuration of their homes – albeit, of course, within the resource constraints of the Soviet command economy. They usually adapted standardised Soviet layouts to meet their specific needs, as will be shown in the next sections.

As a consequence of the described processes, Samarkand's cityscape in 1960 featured a patchwork of different types of private houses, including traditional courtyard houses in the old (pre-colonial) part of the city; residential houses built under Tsarist rule; standardised type houses that were erected on plots issued by the state during the Soviet period; traditional houses built in peri-urban areas on the margins between the city and the surrounding villages/collective farms; and various forms of illicitly built dwellings, often with shanty-town or slum-like character.

---

41 TsGARUz f. 1619, op. 16, d. 4, l.35
42 Smith, Property of Communists, p. 36, 89.
43 Kuznecov, D./Skotnokov, V.: Posobie dlya individual'nogo zastrojshchika, Moscow: Izdatel'stvo ministerstva kommunal'nogo chozyajstva RSFSR 1958.
44 Decree of the CK CPU from 01.06.1962. Available online at http://www.libussr.ru/doc_ussr/usr_5838.htm (accessed 14.06.2019).
45 Petrova, Mariya: Nah am Boden. Privater Hausbau zwischen Wohnungsnot und Landkonflikt im Samarkand der 1950er- und 60er-Jahre, Berlin: De Gruyter 2021.

Despite its specific profile, Samarkand was not a rare exception within the USSR in terms of individual house building. In 1980, about 20 years after the start of mass housing campaign in 1957, the share of *urban* housing space in private hands in the Uzbek SSR remained at 40%, down from 64% in 1960. Figures in other Central Asian SSRs did not differ significantly.[46] A calculation by Tulaganov et al. came to the result that as recently as in the early 2000s, 60%-80% of Central Asian dwellings were made of soil.[47]

SPACE-MAKING IN A COURTYARD HOUSE AND THE AMBIGUOUS TEMPORALITY OF ADOBE BRICKS

The variety of different housing types in Soviet Samarkand, as described above, correlates with the multiple ethnic, linguistic and social backgrounds of the city's residents.[48] In this chapter we concentrate on the courtyard house as a traditional local form that proved its persistence and flexibility throughout the Soviet period. Before examining the practices of building, maintaining and improving private houses in Samarkand it is important to understand the multitude of space-making processes that took place and went far beyond functional considerations.[49] A study on the concept of "remont" in post-Soviet Tajikistan by Wladimir Sgibnev is instructive in this regard. In Tajikistan, *remont,* a term that describes all kinds of repair activities in Russian, is not only about mending something broken. Taking up Lefebvre's theory of "space production", he states that *remont* is rather a culturally embedded creative practice of creating spaces that were perceived as clean, representative and well-kept.[50] In the case of courtyard houses these practices are narrowly connected with local Muslim and Ta-

---

46 Smith, Property of Communists.

47 Tulaganov, A. et al.: "Housing Construction with use of clay materials in Uzbekistan," Kerpic – Living in Earthen Cities, Istanbul: Istanbul Technical University 2005.

48 When looking at the different types of homes it is helpful to differentiate between groups inhabiting Samarkand from the time before the Russian conquest and October Revolution and groups that moved or were moved into the region from Russia and other parts of the Soviet Union, including Russians, Armenians, Ukrainians, Koreans, Crimean Tatars and many others.

49 See Sgibnev, Wladimir: "Remont. Housing Adaptation as Meaningful Practice of Space Production in post-Soviet Tajikistan", in: Europa Regional 22 (2014), p. 53–64, here p. 56.

50 Ibid., p. 55.

jik/Uzbek ideas and ethics of family and household organisation.[51] The form described below is to be understood as an ideal type as it was presented to us by different respondents. It could not always be achieved because of individual, financial or spatial constraints, but it would be aspired to by dwellers.

First, the house had to provide adequate space for different generations usually living together in a household: ideally, grandparents, the families of married sons (the youngest son was expected to stay with his parents permanently and other sons stayed until they had built their own houses) and unmarried daughters (married daughters moved to their husband's family) would have a separate space on their own which was sufficiently heated in winter. The number of stoves in the house was often a good indicator of the number of family "units" living in the house: "we had three stoves, one in our room, the other by grandma and grandpa, and the third was for the brother (who was married, authors)".[52] From spring to autumn, most social life of the household would take place in the courtyard, where typically an *aiwan* provided for a shaded place and served as a rain shelter. An *aiwan* is the centrepiece of every Uzbek courtyard house, a mixture between a roofed pergola and a terrace on the side of one of the courtyard walls, carried by one or more columns. Functional spaces such as toilets, kitchens or washing rooms were sometimes integrated into the main building but were mostly located in separate buildings on the premises. Larger courtyards were often also used to grow fruit and vegetables or keep livestock on the plot, the size of which could range from 200m^2 in the city centre to 2,000m^2 in peri-urban areas.[53]

Equally important, if not more so, was the representative function of a house. In almost every house, one room had to provide enough space or was nearly exclusively reserved for receiving guests and for festivities.[54] When deciding upon the layout of the house, the owner therefore had to find a balance between functionality and representativeness. Cultural norms required a strict separation between the well-maintained and decorated representative part of the house and an area for messy everyday activities which had to be concealed from the view of visitors – especially since these two areas were highly gendered. In households

---

51 For a comprehensive overview of current research on Central Asian families see Roche, Sophie (ed.): The Family in Central Asia: New Perspectives, Berlin: Klaus Schwarz 2017.

52 Interview 7, 25.09.2018, Samarkand, central mahalla.

53 Ibid.; Petrova, Nah am Boden.

54 See Interview 6, 22.09.2018, Samarkand, peri-urban area; also on Islamic festivities, for example the Kurbon celebration.

following Islamic traditions, especially before and during the early Soviet rule, houses were often divided into "female" and "male" – in Uzbek *ičkari* ("inner") and *taškari* ("outer") realms.[55] Other social conventions required periodic changes in the spatial arrangement of the house. While it was common, for example, for older sons to move out and establish their own household once they married, the youngest son had to stay on the premises with his family to take care of his parents, and hence needed separate rooms at the latest by the time of his marriage.

These requirements explain why only very few residents of the Old City were willing to move to the state-built micro-districts even if they received an apartment there. Although some of our respondents recall their admiration for the amenities provided in these apartments, none of them remembers actually wanting to live there.[56] Like the apartments in the micro-districts, the spatial configuration of Soviet standardized designs (*tipovoj proekt*) for individual house building was largely incompatible with the spatial demands of an Uzbek family. These designs included standardised, detached single-family houses of two to four rooms with a square-shaped (c. 10x10m) or rectangular (c. 8x10m) layout. Once a family had managed to receive a plot, it had to request permission from the municipal planning authority to build a *tipovoj proekt*, or more often than not was assigned one. In comparison with compact type houses, the open layout of traditional courtyard houses allowed much more flexibility for setting up and (re)arranging the different spatial elements of an Uzbek or Tajik household at a relatively low cost.

As most rooms were accessible through the courtyard and not through corridors or hallways, additional rooms, extensions or small buildings could successively be added, moved or repurposed.[57] Although the average living space available per capita in the Uzbek SSR (traditional and modern housing taken together) remained relatively low compared with the Soviet Union as a whole,[58] the multifunctional use of the courtyard and relocation of many household activities into the courtyard during the warm season could partly compensate for

---

55 Voronina, V. L.: Narodnye traditsii arkhitektury Uzbekistana, Moscow: Gosudar-stvennoe izdatel'stvo arkhitektury i gradostroitel'stva 1951, p. 12.

56 Interview 6; Interview 7; Interview 2, 17.09.2018, Samarkand, central mahalla.

57 See Sgibnev, Remont, p. 58.

58 E.g. in 1975 the average living space per capita in Samarkand was 5.5 m², SamOGA f. 1658 (Department of Planning and Architecture/Arkhitekturno-planirovochnoe up-ravlenie), op. 1, d. 220, l. 35. The sanitary norm was 9 m² per capita.

this scarcity of indoor living space.[59] For these reasons, many house owners successively transformed their type houses into courtyard houses. This process can be traced not only in interviews with house owners but also in archival documents as people applied for permission to extend their houses or registered extensions retroactively.

This practice, of course, was contrary to the social engineering visions of Soviet architects and planners. Drawing on the sociocultural evolutionary theory of Marx and Engels they assumed that as a result of the increasing urbanisation and "societal development" of Central Asian societies, so-called "extended" families would automatically disappear – the majority of flats in multi-storey apartments were hence planned for the size of "European" nuclear families with 1-3 rooms.[60] By the mid-1970s some voices were arguing for the recognition of traditional housing types in city planning and architecture, but attempts to implement these ideas went no further than the experimental stage.[61]

The extended family remained a widespread type of household in Samarkand throughout the entire Soviet period.[62] Our observations suggest that there is a link between the persistence of social practices of space-making within these extended families and the temporality of adobe brick construction. The latter can be described as ambiguous: The long life of adobe brick construction as a technology that has not fundamentally changed for millennia in Central Asia contrasts with the impermanence of its materiality.[63] The material is vulnerable to weather effects and necessitates a yearly routine of minor repairs after winter and a complete overhaul at least once per generation.[64] One interviewee recalled that her house, passed down from her grandfather to her father, was in a bad state of disrepair because her father, a teacher, lacked the financial resources to

---

59 Rywkin, Michael: "Housing in Central Asia: Demography, Ownership, Tradition. The Uzbek Example", in: Grant, Steven A. (ed.): Soviet Housing and Urban Design, Washington, DC: U.S. Department of Housing and Urban Development 1980, p. 39–43.

60 Rusanova, L. N.: "Demografiya i zhilishche. Po materialam issledovaniya naseleniya Samarkanda", in: Stroitel'stvo i arkhitektura Uzbekistana 3 (1968), p. 28–30.

61 Chebotareva, Z. N.: "V zashchitu plotnoj maloetazhnoj zastrojki", in: Stroitel'stvo i arkhitektura Uzbekistana 1 (1974), p. 28–32.

62 According to the SU-wide census in 1989 an average family in Uzbekistan consisted of 5.5 people. Families consisting of six people or more made up a share of 39%. Boldyrev, V. A.: Itogi perepisi naseleniya SSSR: Naselenie SSSR. Po dannym vsesoyuznoj perepisi naseleniya 1989 g., Moscow: Finansy i statistika 1990, p. 32.

63 As an archaeologist that accompanied us on our field trip reassured us.

64 Interview 2; Interview 8, 26.09.2018, Samarkand, private residential area.

maintain it properly.[65] Repair cycles were often defined by the social temporality of the courtyard house rather than functional considerations. Houses (or at least the parts that would be seen by visitors) had to be in shape for festive events that accompanied life cycle rituals such as circumcisions, weddings or funerals.[66] The cost of renovation could actually account for a considerable part of the total expenses for such events. The material constitution of courtyard houses was hence in constant flux and intimately tied to the natural seasons and the life cycle of their inhabitants.

The recurring cycles of renovating, rebuilding and extending courtyard houses required construction materials that were affordable and readily available. While some industrially manufactured materials such as cement and gravel were available at affordable prices in state warehouses, most of the fundamental materials such as burnt bricks or wood were in constant shortage and reserved for government construction projects. In January 1967, for example, an order by the regional party committee prohibited the use of burnt bricks for building foundations, basement walls, retaining walls, etc. for individually built houses.[67] Even if burnt bricks were available to buy, their costs were prohibitive. According to one respondent, in the 1970s the price for 1,000 bricks was around 120 rubles, which was the average monthly salary of an engineer. At the same time the number of bricks a person was allowed to buy in a year was limited to 2,000. As the construction of a house required between 12,000 and 16,000 bricks, a house made of fired bricks was almost impossible to build legally.[68] Adobe, in contrast, could be sourced from the ground, bypassing the formal Soviet command economy with its shortage of products and freely available professional labour. It was affordable as well – 1,000 pieces would cost 14-15 rubles.[69]

Moreover, when used in accordance with traditional architectural principles for building courtyard houses, adobe bricks offered good insulation properties. The specific orientation of the different elements of courtyard houses, their thick adobe brick walls, their ventilation system and their shady courtyards that usually featured a small garden or a water basin provided for a favourable microclimate and indoor temperature in the summer. While most urban planners in Rus-

---

65 Interview 2.
66 According to the depictions of two interviewees (Interview 2, Interview 6).
67 SamOGA f.1617 (Department for Construction and Architecture of the Provincial Executive Committee/Otdel po delam stroitel'stva i arkhitektury Oblispolkoma), op. 1, d. 233, l. 3.
68 Interview 3, 18.09.2018, Samarkand, private residential.
69 Ibid.

sia dismissed the traditional *mahallas* as being unsanitary and inefficient, those visitors to Central Asia who actually entered courtyard houses could hardly ignore these advantages. Leonid Volynskii, a Russian traveller to Uzbekistan in 1961 who published his experiences in an article in the journal *Novyi Mir*, realised that in summer, the temperature in the traditional mud-brick homes in Tashkent was four or five degrees cooler than elsewhere in the city – including the newly built Soviet model district of Chilanzar, for which Volynskii had only harsh criticism.[70] The labour intensiveness of clay and its low durability as material, however, required constant engagement with house construction, improvement and maintenance and thus brought about a particular form of social mobilisation.

BUILDING IDENTITY WITH CLAY? SELF-HELP PRACTICES IN HOUSE CONSTRUCTION AND THEIR INTERPRETATIONS

While adobe was abundantly available in the ground in Samarkand, using it for house construction was a time- and labour-intensive process. In the absence of a formal open market for the labour and services required, individual house building in Samarkand can be understood in terms of an elaborate strategy of self-help and labour mobilisation. We will now turn to the construction process in more detail. The timing of construction was largely determined by the seasons. The production of bricks was usually scheduled for May, June or July at the latest, when no more rain was to be expected, and the subsequent construction would be done in August-September.[71] Adobe bricks need a special soil, so an experienced moulder would inspect the plot in search of a fitting layer of smooth loess soil, removing the upper metre of ground.[72] To save transportation costs, clay for the bricks was usually extracted either from plot itself, thus at the same time excavating a hole for a foundation or basement, or from nearby.

Skilled workers for technically advanced tasks like foundation or carpeting were often invited or recruited from the pool of relatives. In general, most of the

---

70 See Stronski, Tashkent, p. 228.

71 Interview 6.

72 Samarkand is famous for having very thick (up to 25 m) loess ground, which makes high-rise construction particularly difficult but provides perfect material conditions for earth construction (Interview 1, 19.06.2018, Samarkand, colonial city/microdistrict).

knowledge associated with construction in the *mahallas* and peri-urban region was situated in the realm of traditional building practices rather than in the Soviet system of professional training.[73] Most families would turn to masters of traditional building. One respondent recalled that her sister's husband was an *usto* – a master (in this context a foreman) who could advise them on quantities of building materials to be bought and supervise the construction: "My sister's husband was a technician, he told us how much cement and how many cars of gravel we needed to buy, then how many beams and bricks were needed. And the adobe bricks, we made them here in the yard."[74]

Bricks were made by brigades of two or three. Often the moulders were young men – teenage schoolboys or university students who made bricks as a holiday job.[75] They began by soaking the clay with water and then mixing and stamping it to remove as many air pockets as possible and create a consistent mass. The mass was then left to rest and be stamped again. Bricks were formed with a wooden mould and left to dry in direct sunlight, first on a plain surface and then piled up in small pyramids of three by three pieces. The bricks needed to dry for around 10-15 days to allow more air to escape.[76] According to one of our respondents, a brigade could produce up to 1,000 bricks a day. The hole for the foundation was filled with stones, burnt bricks[77] and, after the mid-1960s, when industrially produced cement became more easily available, also concrete. The bricks would be bound with liquid clay and for the plaster a mixture of clay and straw called *saman* was used. Another traditional technology widely used in Samarkand is called *čub-kori* – a Central Asian version of timber frame with mud-brick infill.[78]

The completion of a house within the period of just one summer was a rare exception. In most cases, people first built a foundation, one or two rooms and the roof, although the whole construction could last several years. One respondent recalled that when he was a schoolboy, his father started building an additional house (two rooms and a kitchen) on the opposite wall of the courtyard for his son's future family, apparently aware of a lengthy process to come. The construction went on to last for eight years. In the case of our respondent it was not

---

73 Fodde, Enrico: "Traditional Earthen Building Techniques in Central Asia", in: International Journal of Architectural Heritage 3 (2009), p. 145–168, here p. 152.

74 Interview 6.

75 Interview 3.

76 Interview 6.

77 Despite the restrictions some people reportedly did manage to obtain burnt bricks, but these practices were always connected with the informal realm of the Soviet economy.

78 Fodde, "Traditional Earthen Building Techniques", p. 149–151.

money that was the limiting factor; instead it was the scarcity of industrially manufactured material: "We were building for so long, because we were waiting for the material. How it went: if you have obtained the beams – wait for boards. Once you've done the floor – wait for the laths. And so we did, room after room."[79]

Whether building a new house or extending or rebuilding a house on an existing plot, self-help construction was usually not a one-time activity; it was a recurrent part of house owners' everyday lives over many years, often with the help of some of their neighbours and relatives. After all, building with clay was a labour-intensive process. As well as hiring day labourers on the informal labour market, the bulk of this manpower was recruited from the wider circle of relatives or neighbours. When referring to these practices of kinship or neighbourhood help, our Tajik and Uzbek respondents used the term *hashar*. A female respondent recalls the – highly gendered – process of collective building as follows: "We would buy wood and roofing slate, but concrete and all this – always by the method of *hashar*. We have a big family, we would invite brothers and nephews to come, prepare one big meal, and they would work."[80]

Originally, the term *hashar* referred to a long-established practice of mobilising a large amount of (more or less) voluntary collective labour for the (re-)construction and cleaning of irrigation canals, which were vital for the oasis and riverside villages and towns in Central Asia.[81] Later it was also used for the collective construction and maintenance of public buildings, like mosques and private residential houses, as well as for seasonal work and activities associated with big festivities.[82] *Hashar* had an important social meaning, as it was based on the principle of reciprocity whereby the exchange of obligations tightened social and informal institutions, thereby producing cultural identities.[83] During the Soviet period, the term was ideologically supported and praised as a local form of socialist collectiveness, in particular in the context of individual housing construction.[84] The idea of collective work for a common purpose fitted perfectly into the ideological framework of Soviet collective practices, such as *subbotnik* –

---

79 Interview 3.
80 Interview 6. See also Interview 7.
81 Obertreis, Imperial Desert Dreams, p. 31.
82 Sehring, Jennifer: The Politics of Water Institutional Reform in Neo-Patrimonial States: A Comparative Analysis of Kyrgyzstan and Tajikistan, Wiesbaden: Verlag für Sozialwissenschaften 2009, p. 70.
83 We are grateful to Sergei Abashin for this remark.
84 Razykov, A.: "Individual'nomu stroitel'stvu zhil'ya povsednevnoe vnimanie", in: Bloknot agitatora Tashkent 3 (1958), p. 11–18.

days of "volunteer", unpaid work on Saturdays –, thereby providing an additional layer to its meaning.[85]

While practices of collective building of houses can be found all around the world, the complexity and persistence of *hashar* suggests that the mobilisation and subjectivation process they entail are specific to Soviet Central Asia.[86] This observation becomes more evident when examining another metaphor that was voiced by different respondents in several variations such as "being close to the ground" (in Russian: *blizko k zemle*) or "living on the ground" (*žit' na zemle*). The expression can generally refer to a down-to-earth person as well as an ability to master one's immediate material environment. People who were "close to the ground" or worked in professions that were "close to the ground", such as masons, carpenters, construction workers and the like, were considered to have the necessary skills to construct and renovate a house – a social categorisation.[87] The expression would be also used to describe a desired living condition (to live on the earth, to have one's own plot) but as a means of cultural delimitation of one's own group. Asked if she ever considered moving to a Soviet micro-district, one of our respondents, who defined herself as Tajik, answered: "No, it was never like this, we are not inclined to migrate, we live on the ground. I cannot imagine myself living on the 16th floor. The people don't have it, we don't have it in our blood, we have been sedentary tribes."[88]

The obviously stylised self-image of the respondent is an example of the narratives that are present in identity discourses in today's Uzbekistan.[89] They con-

---

85 The most known form of "voluntary-compulsory" collective work in Soviet Union was the *subbotniki* – state-organised collective activities for cleaning and tidying public places like streets and parks. In the context of Uzbek Republic the term *hashar* has been "translated" into the logic of collectivisation. For the use of *hashar* during the soviet period and its present connotation in Samarkand see Marteau d'Autry, Christilla: "*Vyjdem vse, kak odin!* 'Allons-y tous comme un seul homme !' Ethnographie d'un *hashar* national dans un quartier de Samarkand, Ouzbékistan", in: Cahiers d'Asie centrale 19/20 (2011), p. 279–301.

86 For other parts of the world see for example Holston's study on *autoconstruction* in Brazil. He argues that self-help construction of private houses can be understood as an "arena of [...] spatial, political, and symbolic mobilizations". Holston, James: "Auto-construction in Working-Class Brazil", in: Cultural Anthropology 6 (1991), p. 447–465, here p. 447.

87 Petrova, Nah am Boden.

88 Interview 2.

89 The issue of Tajik and Uzbek self-images, the urban-sedentary dichotomies and corresponding narratives is part of the debate on Soviet and post-Soviet politics of national-

trast this mode of housing with the idea of living *high above the ground* in the micro-districts. In this regard, answers from all our respondents living in the *mahallas*, Soviet-era private housing districts or semi-urban areas – notwithstanding their ethnic background – were similar. This reminds us to see private house construction as not solely a cultural but also a social phenomenon, one which aside from the desire to live "on the ground" required material and social means as well as practical skills.[90] Those who had neither of the two would rather queue for flats in *mikrorayons*.

The discourse can also be traced in archival sources. The metaphor of being close to the ground, for example, was used by A. Kogan, the director of Samarkand's architectural and planning authority, in a critical memo on the prospects of housing construction in Samarkand in 1966. In the memo, he deplores the long-standing practice of poorly coordinated and fragmentary construction of individual houses that allowed for new mass housing projects only outside the former city limits and at the expense of transferred *kolkhoz* lands. Individual house construction in a city, he wrote, was "morally outdated", all the more so since it implied much higher costs for infrastructure provision than denser forms of housing.[91] Kogan argued for a ban on individual house construction in Samarkand, but at the same time acknowledged that "the inclination of the local population to the ground" would pose a major obstacle for such an initiative.[92]

While the construction of individual houses is often about the exploitation and re-production of non-state spaces, it also appears to be an arena where different cultural and social identities are negotiated, especially notions of being Tajik/Uzbek and "Soviet". The question remains as to whether collective building practices can be generally interpreted as acts of "Soviet" collectivism or acts of (ethno)cultural self-affirmation within the Soviet state. Was this a contradiction at all? Following Abashin's argument about the mix of multiple sub-spaces and identities that often can be found in Central Asia, we suggest in our next sec-

---

ity and identity. Processes like national delimitation in the 1920s, the subsequent establishment of fixed ethnic categories, and nation-building after the breakdown of the Soviet Union have all inevitably left their mark on the identities of Central Asian inhabitants. For an in-depth discussion see Hirsh, Francine: Empire of Nations; Abashin, Sergei: Natsionalizmy v Srednej Azii v poiskakh identichnosti, St. Petersburg: Ateleya 2007.

90 Petrova, Nah am Boden.

91 "Morally outdated" is a Sovietism that can be translated in this context as "not state of the art any more".

92 SamOGA f. 1658, op. 2, d. 146, l. 6–8.

tion that both identities, at times, entered into a mutually stimulating relationship.[93]

TRANSCENDING BOUNDARIES: COMPOSITE HOUSES, HYBRID *MAHALLAS* AND PRAGMATIC STATE OFFICIALS

The (partly retrospective) interpretations of a collective house-building culture among both residents and the state bureaucracy seem to be very much structured along the dichotomies of tradition vs. modernity, low-rise vs. high-rise, urban sedentary vs. nomadic tradition, continuity vs. change, Soviet state bureaucracy vs. people. This was much less the case for actual practices on different levels and their material outcomes. On the contrary, when it comes to the more mundane level of everyday practice our study shows a high level of pragmatism and openness towards different tastes, needs and materials. While the adobe brick walls hardly changed at all, the houses' interiors bore witness to the interaction and mutual influence of notions of *Sovietness* and the patchwork of cultural identities that is characteristic for Samarkand – the material outcome can be described as hybrid.

In one interview, a woman who introduced herself as Tajik and lives in a *mahalla* in the Old City recalls an attitude she and her family internalised in the 1970s towards the traditional architectural elements of their *aiwan*. The *aiwan* of her family dated back to the end of the 19th century, when her great-grandfather, a wealthy merchant, lived in the house. It featured an elaborately carved wooden column and crossbar, and inside it was lavishly decorated with niches and traditional ornaments. Yet in the eyes of the family and their neighbours, "it meant poverty, for them it signified the past".[94] The respondent recalled painting the *aiwan* together with her sister with a light blue paint that was mass-produced by a state company and had come to be used to paint doors and windows up to the remotest village in the Soviet Union. The painted *aiwan* would remain a visible signifier of Soviet material culture within the courtyard house until the late 1980s, providing a contrast to the adobe brick walls of the house's front part.[95] Another interviewee in a different *mahalla* recalls his father covering the ceiling decoration and the carved columns with plywood "because it was practical for

---

93 Abashin, Sovetskii kishlak, p. 610.
94 Interview 2.
95 Interview 2. Traces of the blue paint were still visible at the time of the interview.

craftsmen, for painters".[96] In light of the precarious legal and material conditions for house improvement, pragmatism often prevailed over traditionalism. As for the roofs, few people had any qualms about letting go of clay roofing, which often leaked and required annual repair, when mass-produced fibre cement became a viable alternative for roof covering in the late 1960s and 1970s.[97]

Along with industrially manufactured materials and products, "European" tastes associated with Soviet material culture also began to influence ideas about interior design. While it was very common in Central Asia to put rugs on the floor for sitting and resting – a practice that also contributed to the freeing up of interior living space –, "European" seating furniture became an object of desire and a status symbol. The respondent recalled her envy towards their better-off neighbours who possessed a chandelier and a sofa – "things that our grandparents never had in their interior".[98]

The interior of houses and the appearance of *mahallas* became increasingly characterised by a materiality that can be interpreted as a hybrid between what were commonly understood as traditional and industrial technologies. The connection of the old *mahallas* in the centre to some of the networked urban infrastructures, with electricity taking off in the 1950s and gas arriving in the 1960s,[99] was not only one of the first interventions by the state; it also revealed the boundaries of collective self-help building as it required expertise obtained in the Soviet vocational training system. When his *mahalla* was connected to the gas network in the 1970s, a respondent recalls that:

"We would do everything by ourselves back at that time, we had no need for technicians. Technicians became necessary when you installed gas. [...] We had here a neighbour, he returned from his military service with a Russian wife, and their sons were technicians in our *mahalla* and built stoves in every house."[100]

Many of the old cast-iron stoves that used coal or wood, however, would not be discarded but refurbished to work with gas. Rather than completely replacing older technologies, the arrival of networked infrastructures led to their modification or added another layer of technology.

---

96   Interview 9, 27.09.2018, Samarkand, central *mahalla*.
97   Interview 9, Interview 2.
98   Interview 2.
99   TsGARUz f. 1619, op. 16, d. 4, l. 43 According to statistics, 29.1% of privately owned dwellings had electricity in 1950. The share had increased to 83.4% by 1960.
100   Interview 7.

While Soviet urban planning and infrastructure development in the old city centre would alter the surface layers without fundamentally changing their material set-up, the new, private residential neighbourhoods that were built after 1950 were emerging as hybrids from the very beginning. In many cases, though not always, these areas started out as planned neighbourhoods with geometrical street grids, standardised type houses and mostly same-size plots. These areas were ethnically more diverse than the old *mahallas* in the centre, and instead of district mosques featured Soviet-type teahouses as meeting points and contained planned areas for public infrastructure like clubs, kindergartens and cinemas. Within their plots, however, many residents made use of a relatively high degree of freedom to gradually transform their type houses into courtyard houses and add characteristic elements of Central Asian houses such a pergola (*aiwan*) or a carpet-covered backless divan (*takhta*) in the yards – thus making the neighbourhoods increasingly resemble the central *mahallas*.

To do so, people initially benefited from a housing policy that increasingly tended to intertwine individual initiative and state capacity in the mid-1950s. Khrushchev's famous housing decree of 1957 explicitly lauded a method pioneered in Gorky (today Nizhny Novgorod) known as "peoples' construction", where brigades of non-construction workers built housing in their spare time in return for the guarantee of a new home.[101] At times, Soviet officials displayed an astonishing pragmatism and ideological flexibility when it came to the interpretation of local traditions. In a report on the implementation of the central committee's housing policy in Uzbekistan in 1958, the author praised the translation of the Gorky method into *hashar*, using a number of successful examples as illustrations.[102] In the early 1960s, however, Khrushchev's attitude towards individual construction and personal property in the housing economy changed, as the state capacity for construction had increased and the Soviet micro-district became a core element in the renewed vision of the communist future. In Samarkand, too, the city administration put a temporary halt to the expansion of the city at the expense of the surrounding *kolkhozes* and also stopped individual building.[103]

101 Smith, Property of Communists, p. 75 and 163; Harris, Steven E.: Communism on Tomorrow Street: Mass Housing and Everyday Life after Stalin, Washington, DC/Baltimore, MD: Woodrow Wilson Center Press/Johns Hopkins Univ. Press 2013, p. 154–162; Reid, Susan E.: "Makeshift Modernity: DIY, Craft and the Virtuous Homemaker in New Soviet Housing of the 1960s", in: International Journal for History, Culture and Modernity 2 (2014), p. 87–124, here p. 102.

102 Razykov, Individual'nomu stroitel'stvu, p. 17.

103 Petrova, Nah am Boden.

Yet difficulties for the city administration in controlling individual house construction remained after 1960. Sanctions on illicit building activities were a rare exception. Amidst the chronic housing shortage, the city administration could hardly justify the removal of housing space without offering adequate alternatives. On the contrary, the ex-post legalisation of houses or extensions such as kitchens, garages or additional bedrooms was a frequent administrative routine in the work of the city executive councils. In most cases, bureaucrats displayed an astonishing pragmatism when it came such requests, as the following typical case shows:

"Comrade Chelebiev, when constructing his house, deviated from the approved plan and has built three rooms with total area of 52.20 m^2 instead of 40 m^2 and three additional kitchens. At the current moment he lives there with three families, ten people in total. [...] Considering the fact that the extension of the living area was made in order to provide adequate living conditions for ten people and the total area does not exceed the prescribed norm, the executive committee of the city council has decided to confirm the changes."[104]

Whether on the level of the household, the *mahalla* or the cityscape, when it came to the more mundane questions of managing urban housing, the encounter of people, ideas and materials from both sides of the dichotomies mentioned at the beginning of this section can often be described in terms of coexistence or mutual influence rather than open conflict. The lack of open conflict, however, does not mean that people were not being subjected to state coercion. After all, although they were allowed to build their own houses, many were doing so because their previous dwellings – sometimes less than a kilometre away – had been demolished to make way for state construction projects.[105]

CONCLUSIONS

The starting point for this chapter was the persistence of both clay as a building material and the courtyard house as a type of housing in Samarkand throughout a period when the Soviet leadership under Khrushchev massively stepped up its efforts to bring its vision of socialist housing to fruition. Our account shows how, not only in the old *mahallas* in the city centre but also in newly emerging private residential areas, people maintained traditional collective practices of

---

104 SamOGA f. 1658, op. 2, d. 109, l. 39.
105 Petrova, Nah am Boden.

house building and improvement that offered a high degree of independence from the Soviet housing economy and the state's social engineering attempts.

By building with clay people could capitalise on existing traditional local knowledge and skills associated with this long-established technique. Given the widespread shortage of industrially produced construction material in a part of the Soviet Union that was generally less industrial than central Russia, it was a cost-effective (and in most cases the only) method to meet the requirements for the social production of space in housing. These processes of space-making were largely determined by the persistence of norms and conventions engrained in local traditions of household and family organisation. The physical characteristics of clay as a building material, in turn, induced and maintained social dynamics that differed markedly from those of other building materials, particularly those which were industrially manufactured. While it enabled people quite literally to build a house out of their own soil, it required the recurrent mobilisation of a relatively large amount of labour.

In this connection, *hashar,* a traditional Central Asian concept (as opposed to Soviet terms like *subbotnik*), appears as a strategy to mobilise a significant portion of the required labour among family and neighbours and thus enable a de facto withdrawal from the reach of urban planners. The deliberate reference to *hashar*, along with the projection of the general Russian metaphor of "being close to the ground" on the traditional urban culture of Samarkand, by both private house owners and city officials, set these practices of self-help construction against the Soviet urban modernisation agenda after the late 1950s. This conclusion, of course, needs to be made with due awareness of the construction aspects in oral history accounts, not least those which Hobsbawm has termed the "invention of tradition".[106] Our observations add to a picture of housing as an arena where multiple cultural and social identities were negotiated. At a time when Khrushchev's mass housing campaign, with its focus on multi-storey, prefabricated concrete buildings, began to take off, building courtyard houses with clay inevitably became problematic for city planners and caused tensions.

While these tensions can certainly be identified in the prevalent discourse, we argue that there was a high degree of pragmatism in terms of both practices and policy. Although few people were willing to compromise on the courtyard

---

106 Hobsbawm, Eric J. (ed.): The Invention of Tradition, Cambridge: Cambridge University Press 1999; See also the discussion paper by Dadabaev, Timur: "The Role and Place of Oral History in Central Asian Studies", UI Papers 13, Elliot School of International Affairs, Washington University/CIDOB Center for International Affairs, Barcelona, 2014.

house as a form of housing, they readily adopted industrially manufactured building materials when they became available, together with some "European" trends in interior design. Persistence, it appears, cannot be equated with stagnation. Faced with a chronic housing shortage that took much longer to mitigate than in other parts of the USSR, Soviet officials also displayed a high degree of pragmatism, for example when it came to the ex-post legalisation of individual buildings or their extensions.

Taking up Abashin's metaphor of a *mosaic of Sovietness* for areas of everyday life, we propose the concept of *hybridity* to describe the material outcomes of these processes of mutual influence. In the mundane courtyard house-turned-type house of Samarkand, "Soviet" and "traditional" Central Asian elements merged into novel and original material assemblages. By focusing on the area of house construction, this chapter adds a piece to the jigsaw of the "Soviet experience" in Central Asia.

Maintaining the Mobility of Motor Cars:

The Case of (West) Germany, 1918–1980

Stefan Krebs

Automobiles are not easy, ready-to-use commodities. Motor cars cannot just be bought, driven and eventually sold or scrapped; they are in constant need of regular servicing and, in the event of a breakdown, of repairs in order to preserve their functionality as automobiles. In short, maintenance and repair are "what keeps automobility going".[1] This was particularly the case in the first decades of mass motorisation, when automotive technology was especially prone to failure, but even today, maintenance and repair are recurring moments in the consumption history of the motor car. However, car repair and repair in general are still largely neglected topics in the history of consumption and technology.[2]

This chapter will use the case of Germany to investigate maintenance and repair as a central part of automobility.[3] From the perspective of a history of consumption, "mobility" has always been (and still is) at the heart of car consumption.[4] It is worth noting here that in his book "Short History of the Consumer Society" Wolfgang König described car consumption in the chapter "Mobility

---

1 Graham, Stephen/Thrift, Nigel: "Out of Order: Understanding Repair and Maintenance", in: Theory, Culture & Society 24, 3 (2007), p. 1–25, here p. 15.

2 See the next section for a more detailed discussion of repair and car repair in particular in the history of technology and consumption.

3 For the post-war period, I will focus on West Germany. In East Germany, cars did not become a mass consumer product and professional and self-repair practices differed substantially. See Möser, Kurt: "Thesen zum Pflegen und Reparieren in den Automobilkulturen am Beispiel der DDR", in: Technikgeschichte 79, 3 (2012), p. 207–226.

4 For some car collectors and hobbyist repairers, mobility is not an essential part of consumption practice, but for most ordinary motorists, driving from A to B for professional or leisure purposes is at the heart of automobility.

and Mass Tourism".[5] A motor car loses its *use value* as a consumer item, at least temporarily, in the event of a breakdown,[6] and the *exchange value* of a broken car decreases when it is sold. So maintenance and repair to prevent or remedy malfunctions have been, and still are, necessary and recurrent moments in the consumption of an automobile. However, car repair also became a leisure activity already in the interwar period and especially in the post-war period, when members of the lower classes started to own automobiles. Self-repair was seen as cheaper than taking a car to a professional garage. Furthermore, self-repair also served as a means to shape and foster male identities as skilled and knowledgeable amateur mechanics, although it was still tied to the idea of restoring the mobility function of motor cars. The chapter will look at these two sides of car repair as the need to maintain the mobility function and the practice of a hobbyist consumer activity that promised status, community and identity.

I will largely draw on published sources from repair manuals, trade journals and consumer magazines to show how and why German motorists either repaired their cars themselves or relied on professional repair services from the beginning of mass motorisation in the 1920s to the more widespread use of car electronics at the end of the 1970s.[7] To give an overview of the development of car repair in Germany, I will investigate four aspects that framed repair as a necessary part of car consumption: the (un-)reliability of automobile technology; the emergence of a car repair infrastructure; repair costs, which determined to a large extent whether one could afford to drive a car; and DIY repair practices.[8] I will argue that the success and widespread adoption of automobiles as consumer products was closely tied to the availability of affordable repair services, and that the emergence of a large-scale car repair infrastructure was a prerequisite for mass motorisation. Kevin Borg has situated car mechanics as occupying a middle ground of technology as they neither produce nor own the cars they repair.[9]

---

5   König, Wolfgang: Kleine Geschichte der Konsumgesellschaft. Konsum als Lebensform der Moderne, Stuttgart: Franz Steiner 2008.

6   An exception is old cars; see Lucsko, David N.: "'Proof of Life' – Restoration and Old-Car Patina" (this volume).

7   Electronics had a profound impact on car repair as it removed car technology from the realm of motorists and mechanics. See Krebs, Stefan: "'Dial Gauge versus Senses 1-0': German Auto Mechanics and the Introduction of New Diagnostic Equipment, 1950–1980", in: Technology and Culture 55 (2014), p. 354–389.

8   There are many more aspects of car repair that would be worth investigating, like questions of warranty and goodwill or the issue of faulty repairs.

9   Borg, Kevin L.: Auto Mechanics: Technology and Expertise in Twentieth-Century America, Baltimore: Johns Hopkins University Press 2007, p. 1–12.

However, I will show in this chapter that car repair should in fact be conceptualised as an integral part of car consumption.

CAR REPAIR IN THE HISTORY OF TECHNOLOGY AND CONSUMPTION

Before I start my historical investigation of German car repair culture in more detail, I will briefly look at the still marginalised role of repair and car repair in the history of technology and consumption. Most historians and sociologists of consumption agree that consumption "involv[es] the selection, purchase, use, *maintenance, repair* and disposal of any product or service".[10] In his essay "Beyond Consumerism" Frank Trentmann emphasised that the history of consumption needs to pay more attention to "processes and spaces connected to consumption before and *after* purchase".[11] Furthermore, in "Disruption is Normal" Trentmann looked at the impact of electricity blackouts on everyday life and came to the conclusion that "the more consumption, the more breakdown, tension and patchwork".[12] Despite Trentmann's indisputable achievements in establishing and advancing the history of consumption as a distinct field of historical enquiry, it is telling that maintenance and repair played little to no role in his seminal study "Empire of Things: How We Became a World of Consumers",[13] even though many modern consumer products could only be consumed because maintenance and repair services were offered concurrently – as the example of automobiles will highlight.[14] Only in his chapter "Throwaway Society" did

---

10 Campbell, Colin: "The Sociology of Consumption", in: Miller, Daniel (ed.): Acknowledging Consumption: A Review of New Studies, London/New York: Routledge 1995, p. 95–124, here p. 100 (italics added).

11 Trentmann, Frank: "Beyond Consumerism: New Historical Perspectives on consumption", in: Journal of Contemporary History 39, 3 (2004), p. 373–401, here p. 375 (italics added).

12 Id.: "Disruption is Normal: Blackouts, Breakdowns and the Elasticity of Everyday Life", in: Shove, Elisabeth/Trentmann, Frank/Wilk, Richard (eds.): Time, Consumption, and Everyday Life, Oxford/New York: Berg 2009, p. 67–84, here p. 81.

13 "Maintenance" and "Repair" are not listed in the book's index. See Trentmann, Frank: Empire of Things. How We Became a World of Consumers, from the Fifteenth Century to the Twenty-First, London: Allen Lane 2016.

14 For the importance of automobile repair services see e.g. McIntyre, Stephen: "The Failure of Fordism. Reform of the Automobile Repair Industry, 1913–1940", in: Technology and Culture 41 (2000), p. 269–299.

Trentmann recognise repair as an important element for a more sustainable consumption of technical objects.[15] Furthermore, the history of consumption does not acknowledge that self-repairs were also part of consumption and that they played an important role in shaping consumer identities, as the case of the German motorists will show.

The initial purchase of a consumer object such as a motor car – rather than the actual consumption, i.e. driving the car – is often taken as an indicator of the development of consumer markets and practices. The consumer history of the automobile has often adopted this approach, for example when historian Manuel Schramm used the growing number of automobiles to pin down the advancement of modern consumer society.[16] This purchase-oriented perspective was already being used in the 1920s to forecast the coming automotive demand and the degree of market saturation by looking at the more advanced spread of other consumer products like the telephone.[17] In his "Short History of Consumption", Wolfgang König also drew on sales figures for bicycles, motorcycles and automobiles to describe the emergence of German consumer society but he additionally mentioned leisure travel as a form of automobile consumption. He attributed the initially low automotive density in Germany to the low purchasing power of the middle class, without explicitly mentioning the high after-sales costs.[18] In his essay "The Automobile in Germany", König provided further details, mentioning that annual maintenance costs could represent up to 50% of the initial purchase price of a car.[19] For Richard Vahrenkamp the consolidation of the sociotechnical system of the automobile in Germany, including the ever-denser network of petrol stations, roadside assistance, dealers and specialised repair shops, was a precondition for the gradual emergence of mass motorisation in the late 1920s.[20] German car manufacturers and service providers adopted the American

15 Trentmann, Empire, ch. "Throwaway Society".

16 Schramm, Manuel: "Konsumgeschichte", in: Docupedia-Zeitgeschichte, URL: docupedia.de/zg/schramm_konsumgeschichte_v2_de_2012 (accessed 19.03.2020).

17 Buschmann, Johannes: "Vorbedingungen der Verkehrsmotorisierung in Deutschland", in: Allgemeine Automobilzeitung (AAZ) 26, 47a (1925), p. 40–43.

18 König, Konsumgesellschaft, p. 173 and 175.

19 König, Wolfgang: "Das Automobil in Deutschland. Ein Versuch über den homo *automobilis*", in: Reith, Reinhold/Meyer, Torsten (eds.): "Luxus und Konsum" – eine historische Annäherung, Münster et al.: Waxmann 2003, p. 117–128, p. 121; see also McIntyre, "Failure", p. 274.

20 Vahrenkamp, Richard: "Die Rolle von Handel und Dienstleistungen beim Aufbau des 'Systems Automobil' in den 1920er Jahren", in: Vierteljahrschrift für Sozial- und Wirtschaftsgeschichte 103, 4 (2016), p. 428–451; see also Petersen, Sonja: "'... anner

slogan of "customer service" as the key to successful mass consumption of automobility. However, Vahrenkamp did not conceptualise the different services as part and parcel of automobile consumption.

While the repair of consumer products has found little attention in the history of consumption, historians of technology have investigated the topic of repair more closely over the past fifteen years.[21] Car repair has been studied by, among others, Stephen McIntyre, Kevin Borg and the author himself. However, the primary focus of these studies has been car mechanics, professional repair as a socio-technical practice and the mechanic-driver relationship, not maintenance and repair as necessary aspects of automobile consumption.[22] Kathleen Franz has investigated repair practices of US motorists and automobile tinkering as a form of consumption from about 1900 to 1939.[23] She stresses the "mechanical ingenuity among the new generation of motor travelers" and that "to drive and repair the machine became tools that consumers used to articulate their varying agendas for greater spatial and cultural autonomy".[24] Tinkering with automobile technology helped American motorists to construct their consumer identity.[25] Still, Franz did not look at car repair in detail and instead focused on car accessories and modifications and motorists' inventions.

THE UNRELIABILITY OF EARLY AUTOMOTIVE TECHNOLOGY

The early motor car was notoriously unreliable. Wolfgang König quotes automotive pioneer August Horch, who claimed that an automobile had to be repaired about every 100 kilometres.[26] Similar statements from early motorcyclists sug-

Tanke': Tankstellen – ein Forschungsüberblick", in: Technikgeschichte 83 (2016), p. 71–93.

21 For a more detailed overview see Krebs, Stefan/Weber, Heike: "Rethinking the History of Repair" (this volume).

22 Borg, Auto Mechanics; McIntyre, "The Failure"; Krebs, Stefan: "'Sobbing, Whining, Rumbling' – Listening to Automobiles as Social Practice", in: Karin Bijsterveld/Trevor Pinch (eds.): The Oxford Handbook of Sound Studies, Oxford/New York: Oxford University Press 2012, p. 79–101; id., "Dial Gauge".

23 Franz, Kathleen: Tinkering: Consumers Reinvent the Early Automobile, Philadelphia: University of Pennsylvania Press 2005.

24 Ibid., p. 1 and 3.

25 Ibid., p. 11.

26 König, Das Automobil, p. 118.

gest that an hour of driving was followed by another two hours of tinkering.[27] Even if Horch slightly exaggerated the error-proneness of automobiles it was for good reason that early car owners usually employed chauffeur-mechanics: "The chauffeur, a learned mechanic, maintain[ed] the complicated mechanism of the motor car and ke[pt] it ready for a ride."[28] It is also telling that the most popular German handbook for self-driving motorists was called "Without a Chauffeur" (*Ohne Chauffeur*). The author emphasised that a good driver had to acquire not only driving aptitude but also extensive repair knowledge and skills, much like a chauffeur-mechanic.[29]

While only a few well-off German motorists took to the wheel themselves before the First World War, German automotive journalists started a lively discussion after the war about the aspiration to mass motorisation in Germany. Most contributors agreed that the susceptibility to defects was one of the major obstacles that hampered upper-middle-class people from owning an automobile. In other words, increasing the reliability of affordable automobiles was seen as an essential prerequisite for mass motorisation.[30] The well-known automotive journalist Berger von Lengerke stated: "First of all and most importantly, these cars have to be constructed in such a way that they can be driven without a mechanic."[31] He also offered several suggestions as to how to design better cars, e.g. through the replacement of wood by sheet metal for the construction of the body, or the reduction of the number of lubrication points. The latter principle would be more convenient for drivers and offer more robustness.[32] Other trade

---

27 Lützen, Wolf Dieter: "Radfahren, Motorsport, Autobesitz. Motorisierung zwischen Gebrauchswerten und Statuserwerb", in: Ruppert, Wolfgang (ed.): Die Arbeiter: Lebensformen, Alltag und Kultur von der Frühindustrialisierung bis zum "Wirtschaftswunder", Munich: C. H. Beck 1986, p. 367–377, here p. 370. Other consumer products like colour TVs were similarly error-prone in their early days. See Weber, Heike: "Mending or Ending? Consumer Durables and Practices of Reuse, Repair and Disposal in West Germany, 1960s–1980s" (this volume).

28 Friedmann, P.: "Der Kraftwagen des Selbstfahrers", in: AAZ 26, 41 (1925), p. 25; see also the chapter "The Problem with Chauffeur-Mechanics" in: Borg, Auto Mechanics.

29 Filius [Schmal, Adolf]: Ohne Chauffeur: Ein Handbuch für Besitzer von Automobilen und Motorradfahrer, Berlin: Klasing 1919.

30 In the US, the Ford Motor Company desperately tried to improve maintenance and repair services between 1913 and 1925 as a reaction to growing customer dissatisfaction with error-prone cars. See McIntyre, "The Failure".

31 Lengerke, Berger von: "Zeitgemäße Konstruktions-Richtlinien", in: AAZ 20, 15 (1919), p. 11–13, here p. 11.

32 Ibid., p. 13.

authors agreed that motor cars should no longer be elitist sports machines that could not be driven over longer distances without a mechanic on board: "He [the self-driving motorist] should be able to regulate and maintain all mechanical parts, at least those in need during a ride, without getting his hands dirty."[33]

German automotive manufacturers indeed introduced several technical innovations that facilitated operation and maintenance, e.g. electric starters. While in 1918 most cars still had manual starters, by 1922 67.7% of all new cars were equipped with electric starters, and the following year the figure increased to 82.5%.[34] At the same time, in the space of a few years the open car body was replaced by the closed sedan, "Innenlenker" in German, a body type that simplified year-round operation. This development went hand in hand with the substitution of wooden bodies with sheet metal bodies that increased the durability of motor cars.[35]

For the interwar period, it is difficult to find figures on the susceptibility to defects or breakdown frequency of motor cars. While we do not have data from Germany, 1926 figures from the British Royal Automobile Club reveal the most common causes of breakdown, and we can assume that German and British car technology was relatively similar at that time. The numbers came from the club's "get you home service", a tow-away service for broken down cars. As the club had to pay for each operation, an expert had to investigate every case and report the technical cause, and this information was added to the club's own breakdown statistics. The figures, not including accidents, revealed that the main source of problems was the ignition with 22.7%, followed by the rear axle with 12.4%, cylinders and pistons with 6.9%, tyres and suspension with 6.1%, and the drive shaft with 5.1%.[36] The American Automobile Association also published a breakdown of statistics in 1928. The figures came from 14 regional clubs offering roadside assistance. The numbers, based on 32,993 calls from motorists in distress, showed the main source of problems to be the battery and starter, with 22.3%, followed by tyres with 20.7%. The other categories were not comparable with the British survey: 20.3% of engines would not start, 17.2% of cars had to

33 Friedmann, "Der Kraftwagen", p. 25.
34 Dierfeld, Benno R.: "Deutscher Kraftfahrzeugbau im Jahre 1923", in: AAZ 24, 12 (1923), p. 27–29.
35 Krebs, Stefan: "The French Quest for the Silent Car Body: Technology, Comfort and Distinction in the Interwar Period", in: Transfers: Interdisciplinary Journal of Mobility Studies 1, 3 (2011), p. 64–89.
36 Witte: "Welches sind die häufigsten Reparaturen am Kraftwagen?", in: AAZ 27, 33 (1926), p. 16.

be towed away, and 5.9% of problems were attributed to unspecified causes.[37] Unfortunately, the British and American statistics do not include information about the age of the cars or the average mileage.

While we do not have comparable figures from Germany, trade journals frequently discussed the unreliability of current automotive technology and they argued for a design principle that would take repairability more seriously. In particular, parts that had to be readjusted frequently, like valves, should be easily accessible for motorists. This was clearly not always the case, and articles about the new models for the coming season were critical of those that did not prioritise a repair-friendly design. One journalist said: "In my opinion, it would be strongly advisable for car engineers not only to work in the design office but also in practice, with a manufactured car, to try to disassemble what they have designed. This would result in many designs being changed once the responsible designer, hands aching, realises how unnecessarily difficult such a job can be."[38]

IN NEED OF REGULAR REPAIRS:
USER SURVEYS FROM THE 1960s AND 1970s

Although the National Socialists promoted mass motorisation after they seized power, actual developments lagged behind expectations, and the Second World War put all ideas of private automobility and worries about unreliable cars on the backburner. But by around 1950, automobile production and automobility in West Germany had recovered from the effects of the war; they soon returned to the level of the interwar years, and the 1960s and 1970s witnessed unprecedented growth in car production and ownership. During the recovery phase, repairability remained an important topic. Test reports of new models regularly discussed the accessibility of batteries, carburettors and air filters – components that needed regular maintenance and readjustment.[39] Automobile manufacturers also investigated how to increase the reliability of new cars. During the winter of 1955, BMW organised long-distance test drives of several models to document wear and tear. They invited a group of ordinary motorists to drive each car around 30,000 km under a range of challenging weather and driving conditions.

---

37 Anon.: "Amerikanischer Straßendienst für Kraftfahrzeuge", in: AAZ 29, 10 (1928), p. 43–44.

38 Kink: "Moderne Kraftwagen und deren Reparatur", in: AAZ 26, 48b (1925), p. 28.

39 See e.g. Anon.: "Die ADAC Motorwelt fuhr: den Mercedes-Benz Typ 180", in: ADAC Motorwelt 6 (1953), p. 629–631.

Technicians then checked the performance of all the technical components that were known to cause technical problems.[40]

For many motorists, reliability was also an important reason to buy a particular brand or model. In the late 1960s, customers of the new mid-range VW 1500, for example, named in a market survey the following main reasons for purchase: 51% cost effectiveness, 44.3% steadiness, 32.5% robustness and longevity, and 32.2% good customer service.[41]

A more systematic investigation of the actual weaknesses of automobiles started at the end of the 1960s. The club magazine of the ADAC, the largest German automobile club, invited its readers to take part in "Field Test" (*Praxis-Test*) surveys. Owners of specific models were invited to fill out detailed questionnaires about their experience with petrol consumption, the life span of tyres, regular maintenance services and unexpected breakdowns, and in the event of malfunctions to list which parts were affected. This is quite remarkable as it shows that users were systematically involved in defining and evaluating the actual maintenance needs of automobiles. In the first test 1,733 owners of five mid-range car models, e.g. the Ford 17 M and the Renault 16, took part. The cars had an average mileage of 25,813 km. The published results revealed, for example, that drivers of a Renault 16 had to visit a workshop for repairs 1.2 times between regular maintenance services, compared with 2.8 times for drivers of a Fiat 125. In general, the parts most susceptible to malfunction were the engine, carburettor, cooling system, exhaust pipe and clutch. However, the survey revealed considerable differences between car models: 60% of Fiat 125s had problems with the exhaust, compared to 11% of Renault 16s; 32% of Opel Rekords had problems with the clutch compared to only 5% of Ford 17 Ms.[42] From these results, we can deduct that the reliability of the ignition, starter and battery, the components that caused most problems during the interwar period, seems to have increased, as they are no longer listed in the first field test. However, this was only true for mid-range cars, as another survey showed. This time small cars like the popular VW 1200 were tested. While the overall reliability was comparable with that of larger cars – drivers of a Simca 1000 LS had to see a mechanic 1.3 times between services, owners of a Fiat 859 N 2.6 times –, different components caused trouble: the engine, starter, generator, ignition and carburettor.[43]

---

40 K.W.: "Versuchsauftrag 88", in: ADAC Motorwelt 8 (1955), p. 66–67.

41 Anon.: "VW 1500: Schwarzbrot aller Klassen", in: Der Spiegel 50/1968, p. 110.

42 MC: "Leser-Bilanz nach 45 Millionen km", in: ADAC Motorwelt 23 (1970), p. 40–48.

43 Lotz, Heiner: "Nach 40 Millionen km", in: ADAC Motorwelt 24, 11 (1971), p. 60–71.

And 1978 figures from the ADAC roadside service showed yet another list of unreliable car parts: cable connectors, throttle cable, breaker contacts and fan belts were to blame for one in every four breakdowns.[44] So it is difficult to estimate which parts of automotive technology became more reliable over time and which troubled car drivers all the time. Before I conclude this section, it is interesting to note that the 1977 field test of medium-sized cars revealed that a Japanese car, the Toyota Corolla, was significantly more reliable than the other cars tested. On average, Corolla owners had to see a mechanic only 0.5 times between services; no other car tested had a figure below 1 in the 1970s.[45] The reliability of Toyota was also reflected in roadside service figures and the brand became known for its quality.

Susceptibility to breakdown and the frequency of malfunctions significantly influenced the possibility or impossibility of car consumption. Early motor cars were so error-prone that driving without a mechanic on board was difficult. Trade journalists urged manufacturers to produce more reliable cars to convince more customers to buy an automobile, but field tests from the 1960s and 70s show that technical problems were still inherent to car mobility. In the following section I will look at the emergence of a (roadside) repair infrastructure that helped car drivers to cope with unreliable automobiles.

THE EMERGENCE OF AUTOMOBILE REPAIR INFRASTRUCTURE

As seen in the previous section, the unreliability of motor cars jeopardised the consumption of (auto-)mobility – and not only in the early years. When driving, and especially outside larger cities, motorists had to be prepared to deal with smaller and larger malfunctions at any time. Before the First World War, when most automobiles were driven by chauffeurs, a mechanic was always on board, because chauffeur-mechanics were trained to drive and to repair a motor car on the road if necessary. This is also why motor cars were equipped with a large

---

44 Id.: "Diese vier Teile waren an jeder vierten Panne schuld", in: ADAC Motorwelt 31 (1978), p. 10–19.

45 Id.: "Vier Europäer und ein Japaner. Er war der Zuverlässigste", in: ADAC Motorwelt 30, 4 (1977), p. 42–46.

tool set and some spares of parts that were well known for causing problems.[46] At home, the garage was also a workshop where chauffeurs could carry out major repairs in addition to regular maintenance, cleaning and greasing. In many cases, larger rental garages offered a parking space together with the necessary tools, parts and work areas for all kinds of repair work.[47] If the chauffeur could not repair the car himself, it had to be towed to the specialist workshop of a car dealer or manufacturer. Independent professional auto mechanics only gradually emerged out of traditional trades like blacksmiths, fitters and other metalworkers.[48]

In the interwar period, the motoring press praised self-driving as a new ideal. Manufacturers and trade journalists agreed that mass motorisation could only really get going in Germany if members of the middle class took to the wheel, and they had to do so themselves, as they did not possess the financial means to employ a chauffeur. However, this new practice required one of two things: either drivers had slip into the role of chauffeurs and drive and repair their cars themselves, or substantially more maintenance and repair services had to be offered to help drivers with breakdowns on the road. Both options were discussed at length in the motoring press: self-driving motorists were urged to acquire the necessary knowledge and skills to repair their automobiles, and the automobile business was advised to offer more and better services. In 1925, for example, an article in the "General Automobile Newspaper" (*Allgemeine Automobil-Zeitung*, hereafter AAZ) argued not only that road conditions and traffic signs needed to be improved and expanded, but that more service stations and other auxiliary facilities were urgently needed to ease the increase in automobile consumption, e.g. a pick-up and delivery service for workshops or a maintenance subscription service. Furthermore, petrol stations, garages and workshops had to learn to offer a real "service to the customer", as emphasised in the United States. This plea partly came in response to a rise in complaints from motorists about faulty repair jobs and excessive bills.[49]

During the 1920s, the network of professional maintenance and repair businesses in Germany became denser. At the end of the decade, about 20,000 spe-

46   Krebs, "'Notschrei eines Automobilisten' oder die Herausbildung des deutschen Kfz-Handwerks in der Zwischenkriegszeit", in: Technikgeschichte 79, 3 (2012), p. 185–206; see also Borg, Auto Mechanics, ch. "The Problem with Chauffeur-Mechanics".

47   Anon.: "Betriebsorganisation von Großgaragen", in: AAZ 26, 28 (1925), p. 45; Vahrenkamp, "Die Rolle".

48   Krebs, "Notschrei"; Borg, Auto Mechanics.

49   Buschmann: "Vorbedingungen der Verkehrsmotorisierung in Deutschland", in: AAZ 26, 47a (1925), p. 42.

cialised auto repair shops offered their services for a fleet of 420,000 cars.[50] In addition, the automotive trade underwent a process of professionalisation, as I have shown elsewhere,[51] and founded its own trade guilds, with apprenticeship schemes and mandatory exams to improve and secure the qualification level of auto mechanics. After coming to power in 1933, the National Socialists promoted the still stagnating mass motorisation of Germany. To this end, they established a legal framework of compulsory auto mechanics guilds and only permitted repair shops to be opened by qualified master craftsmen. The argument for the introduction of these measures was that the difficult relationship of trust between motorists and mechanics was preventing too many prospective motorists from buying an automobile, and the new system would encourage these people to take the leap and actually purchase their own car.

In addition to the emergence of a network of professional repair shops, automobile clubs and dealer associations started the first roadside repair and tow-away services in the late 1920s.[52] In the autumn of 1927, for example, the ADAC decided to set up and operate a first roadside assistance service the following year. A small fleet of 40 motorcycles, cars and light trucks patrolled main roads to help motorists in distress – all motorists, but club members were given priority. In addition, the ADAC set up a line of telephone boxes along main roads that drivers could use to call for help if they were experiencing technical problems. In contradiction to their announcement that they would foster mass automobility, the National Socialists forced the German automobile clubs to stop their roadside assistance systems in 1933.[53]

After the war, the ADAC decided to re-establish a road service in 1951. As a first step, the motor club established contracts with 3,000 repair shops. The commissioned auto mechanics offered members a free roadside assistance and tow-away service. However, most shops were located close to main roads and motorways.[54] To provide help away from the main road network, the ADAC encouraged skilled members to join the club's volunteer service. These volunteers

---

50 Krebs, "Notschrei", p. 194.

51 Ibid., p. 198–201.

52 Vahrenkamp, "Die Rolle".

53 Anon.: "ADAC-Strassenwacht", in: ADAC Motorwelt 6 (1953), p. 711. From the available sources, it is not clear what motivated that move. In 1933, the National Socialists forced all existing German automobile clubs to merge into the new German Automobile Club (DDAC). The discontinuation of roadside services may have been part of that restructuring process. See Fack, Dietmar: Automobil, Verkehr und Erziehung, Wiesbaden: VS Verlag 2000, p. 326–327.

54 Anon.: "ADAC-Strassendienst", in: ADAC Motorwelt 4 (1951), p. 10.

had a sticker on their cars and promised to help fellow motorists. In 1953, there were about 7,500 such volunteers who reported more than 21,500 cases of assistance;[55] two years later, more than 23,000 volunteers reported some 30,000 repair services.[56] At about the same time, the ADAC re-introduced roadside patrols.[57] Initially, 60 club mechanics with sidecar motorcycles equipped with tools and spare parts patrolled the roads. For the year 1955, the club proudly reported that the roadside mechanics had repaired some 150,000 motor cars.[58] The ADAC gradually expanded its assistance network in the following decades, and other motor clubs offered similar services, too.

In the field of professional auto mechanics, it is noteworthy that the number of independent repair shops diminished during the 1940s.[59] In 1949, some 15,000 workshops with 100,000 mechanics took care of one million vehicles (including commercial vehicles). By 1966, the number of workshops had increased slightly to 18,400, but at the same time the number of cars had soared to twelve million.[60] The lower ratio of repair shops to automobiles led to longer waiting times for a workshop appointment and unsatisfactory service quality. The conviction that bad repair jobs were not only a subjective judgement by motorists led to a large-scale assessment of repair shops by the ADAC in collaboration with the popular magazine *Stern*. In 1970, six specially prepared cars with ten built-in faults were sent to 120 different garages. The idea was that a good regular maintenance service should find and fix all malfunctions such as a burntout headlight bulb. However, on average only 59% of the work was done during the planned maintenance service.[61] Five years later, only one workshop out of another 120 tested fixed all the faults.[62] The workshop tests highlight the important role of automobile clubs as mediators between producers and consumers. The test reports sparked a public debate about poor repair services and led to

---

55 Anon.: "Ausbau der ADAC-Leistungen", in: ADAC Motorwelt 6 (1953), p. 367.
56 Bretz, Hans: "Die Armee des Friedens", in: ADAC Motorwelt 9 (1956), p. 706–707.
57 Anon.: "Die ADAC-Strassenwacht", in: ADAC Motorwelt 6 (1953), p. 617.
58 Bretz, "Die Armee des Friedens".
59 From the available sources it is unclear whether this was an impact of the war or if the sector was consolidated for economic reasons.
60 Anon.: "Reparaturen müssen so teuer sein", in: ADAC Motorwelt 19 (1966), p. 42–48. In 1960, the number of car mechanics was 179,000, in 1970 226,000. Krebs, "Dial Gauge", p. 368.
61 MC: "Von 10 Inspektions-Arbeiten nur 6 gemacht!", in: ADAC Motorwelt 23, 5 (1970), p. 38–48.
62 Caroselli, Manfred: "Diese 6 Wagen waren in 120 Werkstätten zur Inspektion", in: ADAC Motorwelt 28, 4 (1975), p. 4–10.

some changes in the legal framework that made it easier for motorists to check their workshop bills and identify any unsatisfactory or overpriced repairs. Manufacturers drew another conclusion: they saw the main cause of the workshop problems as being the poor ratio of mechanics to automobiles and introduced new tools and devices for the rationalisation of diagnostic and repair work. Volkswagen, for example, introduced a standardised service check with some semi-automatic tests in 1968, followed by a "computer diagnosis" in 1971. The latter was an attempt to fully automate checks on 88 parts and functions to eliminate the error-prone human factor in car diagnostics – with little success. Volkswagen abandoned the computer diagnosis after receiving numerous complaints from customers and mechanics for several years.[63]

To sum up, the establishment and expansion of a repair infrastructure with professional repair shops, tow-away and distress call services and roadside assistance helped more and more people take to the wheel, despite the prevailing unreliability of automobiles. While basic technical knowledge and skills, as well as tools and spare parts, were still a necessity in the interwar years to drive outside larger cities, better infrastructures enabled less knowledgeable and skilled motorists to use their automobiles without worrying too much about a possible breakdown on the road. From the late 1950s onwards, ownership of automobiles in West Germany increased on average by 850,000 cars every year. However, this reassurance came at a price, for example the yearly membership fees for an automobile club. In 1962, ADAC membership cost around 31.70 DM while the monthly average income was 610 DM.[64]

THE DEVELOPMENT OF AUTOMOBILE REPAIR COSTS

While the membership fee for an automobile club seemed reasonable for people with an average income, at least in the 1960s, the cost of maintenance and repair had a more decisive impact on whether people could afford to own a car at all. Although we lack exact figures for the first 50 years of automobile consumption

63 Krebs, Stefan: "Diagnose nach Gehör? Die Aushandlung neuer Wissensformen in der Kfz-Diagnose (1950–1980)", in: Ferrum 86 (2014), p. 79–88; for early attempts to rationalise car repair work see McIntyre, "Failure".

64 Motor clubs in other European countries also offered similar services to motorists, see e.g. Anon.: "Panne im Urlaub?", in: ADAC Motorwelt 26, 6 (1973), p. 136–142. For the average income in 1962 see Bundesgesetzblatt I, Cologne: Bundesanzeiger 2002, p. 869–870.

in Germany it is safe to argue that high maintenance costs hampered mass consumption of cars until the 1950s. This section will take a closer look at the development of these maintenance costs. One problem in doing so is not only the lack of detailed numbers but also the definition of maintenance costs in general. Broadly speaking, maintenance costs include all steady and usage-dependent costs to keep an automobile running, i.e. taxes, rent for a garage, fuel and tyre consumption, costs of technical services and repairs. However, the sources do not always indicate which costs are included or excluded from the numbers given. This can make it particularly difficult to estimate the costs for technical services and repairs that are of interest for this section.

An article in the AAZ stated in 1919 that before the war a self-driving motorist had to pay 28-30 pfennigs per kilometre for a small 5 HP car. With an annual mileage of 10,000 kilometres, this came to about 3,000 RM per year, with such a car costing around 4,500 RM to purchase.[65] So for a small car, maintenance costs amounted to two-thirds of the purchase price, while the average annual income was 2,010 RM.[66] However, the article stated that such a small car "was quite affordable for middle-class white-collar workers".[67] After the First World War, the goal of mass consumption of automobiles did not come to fruition in Germany. One important reason was the declining purchasing power of the middle classes because of the wartime and post-war inflation, which pushed car ownership out of reach.[68] Still, the automobile stock in Germany had doubled in 1922 compared to 1914.[69] However, only a small wealthy class could afford car ownership; it is telling that around 50% of cars at that time were still driven by a chauffeur.[70]

Another article from 1922 revealed that the majority of usage-dependent costs were not attributed to tyre wear or fuel but to buying spare parts to replace worn-out components.[71] The high repair costs sparked a debate about the unsatisfactory relationship between initial purchase and follow-up repair costs. One

65 Lengerke, Berger von: "Kleinautos und kleine Wagen", in: AAZ 20, 40 (1919), p. 19–20.
66 Bundesgesetzblatt I (2002), p. 869–870.
67 Lengerke, "Kleinautos", p. 19.
68 Feldman, Gerald D.: The Great Disorder. Politics, Economics, and Society in the German Inflation, 1914–1924, New York/Oxford: Oxford University Press 1993.
69 Anon.: "Die Automobile der Welt", in: AAZ 23, 51, 52 (1922), p. 28.
70 Anon.: "Die Zählung der Kraftwagen", in: AAZ 23, 23 (1922), p. 31–32; Vahrenkamp, "Die Rolle".
71 Freiherr v. Löw: "Kleine oder große Wagen?", in: AAZ 23, 16 (1922), p. 29.

journalist accused manufacturers of building cheap but repair-unfriendly automobiles:

"Such a car will be cheap. However, because the design only took into consideration the
market price and not use and maintenance it is likely that you will have to remove the
whole engine and the cylinders even if you only want to change a worn out connecting rod
bush. […] So what is the point of producing a cheap car if at the slightest malfunction and
repair the owner starts to hate it?"[72]

In addition to this discussion about construction principles and repairability, we
can learn more about the overall lifespan of motor cars from a survey carried out
in the United States in the mid-1920s. The survey revealed that automobiles lasted only seven years, and that the average car was about three years old.[73] These
figures, even if they might have been slightly different in Germany at the time,
suggest that the repairability of automobiles was rather limited and that increasing maintenance costs made older cars uneconomical within a few years. Or, in
other words, car consumption required the purchase of a factory new car every
three to seven years.[74]

Another type of source, advertisements, also points to the importance of low
maintenance costs for potential buyers. From the 1920s onwards, the economic
aspects of car ownership became an important advertising slogan for manufacturers. In particular, small cars were advertised for their low maintenance costs.
The Wanderer campaign in 1922 praised its "exceptional economy" that enabled
"cheap business travel and commuting".[75] A few years later, Opel claimed that
the new 4 HP model was the "most economical car in the world".[76]

In the post-war period, in particular from the late 1950s, when car ownership
came within reach of the lower middle class and the working class, manufacturers of small and medium-sized cars continued to advertise the low maintenance
costs of their models. In 1965, Volkswagen announced in an advertisement that
"the most economical thing about a Volkswagen is how long it is economical",
and went on to claim that "at VW, even the repairs are economical".[77] It is hard-

72 Wa. O.: "Technischer Komfort bei Kraftfahrzeugen", in: AAZ 24, 40 (1923), p. 38.
73 Anon.: "Die Lebensdauer des Automobils", in: AAZ 27, 28 (1926), p. 26.
74 The short lifespan of cars was also an important factor in explaining why second-hand
markets did not really emerge before the post-war period.
75 Wanderer: advertisement, in: AAZ 23, 7 (1922), p. 37.
76 The 4 HP Opel cost 2,800 RM, with an annual tax of 150 RM. Opel 4 PS: advertisement, in: AAZ 28, 36 (1927), p. 41.
77 VW: advertisement, in: ADAC Motorwelt 18, 1 (1965), p. 31.

ly surprising that market research surveys found that maintenance costs were a key factor for customers of small and mid-range cars. In the case of the VW 1500, for example, 51% of buyers in 1968 pointed to "cost-effectiveness" as the main reason for buying, followed by "reliability" with 44.3%.[78] Furthermore, manufacturers like Ford emphasised that they did not sell "throw-away car[s]"; this was presumably a pre-emptive reaction to the impending debate over planned obsolescence.[79]

From the 1950s on, we have more detailed figures about maintenance costs. Automobile clubs like the ADAC started to regularly publish calculations of various cost items. For 1953, these figures showed that the fixed costs of a VW Standard amounted to 1,994 DM per annum and the running costs per 100 km to 10.68 DM. At the previously assumed mileage of 10,000 km this amounted to 1,068 DM. For repairs only, including spare parts and a reserve for a major overhaul, the ADAC calculated 2.50 DM per 100 km.[80] Considering that the average annual salary was 4,061 DM, it is clear that car ownership was still out of reach for most people.[81] Twenty years later, the calculated repair costs for a VW 1300 had doubled to 5.55 DM per 100 km, but the average salary had more than quadrupled to 18,295 DM.[82] Thus automobile consumption had come within reach of the lower middle class.

In addition to calculated repair costs, we can find figures of actual repair costs from motorists. Starting in 1955, club members were regularly asked to report their repair and maintenance costs to the club journal. These surveys revealed that actual repair costs were initially still slightly higher than the calculated costs. In 1955, for example, owners of cars with a 1,200 cc engine reported to the *ADAC Motorwelt* that they had paid on average 3 DM per 100 km for repairs (calculated on the basis of an annual mileage of 10,000 km).[83] In 1971, a similar survey showed that the average repair costs for a VW 1200 had dropped to 1.10

78 Anon., "VW 1500".

79 Ford: advertisement, in: ADAC Motorwelt 29, 5 (1976), p. 43; on the debate over obsolescence see Weber, Heike: "Made to Break? Lebensdauer, Reparierbarkeit und Obsoleszenz in der Geschichte des Massenkonsums von Technik", in: Krebs, Stefan/Schabacher, Gabriele/Weber, Heike (eds.): Kulturen des Reparierens: Dinge – Wissen – Praktiken, Bielefeld: transcript 2018, p. 49–83.

80 Wa.: "Was mein Wagen kostet", in: ADAC Motorwelt 6 (1953), p. 61–62.

81 Bundesgesetzblatt I (2002), p. 869–870.

82 Anon.: "Autofahren war noch nie so teuer", in: ADAC Motorwelt 26, 4 (1973), p. 70–73; Bundesgesetzblatt I (2002), p. 869–870.

83 80,000 members took part in the survey. Anon.: "Ihr Wagen in der Statistik", in: ADAC Motorwelt 9 (1956), p. 584–585.

DM per 100 km, which is only one fifth of the calculated repair costs, as shown above.[84] It is difficult to interpret the different figures because the sources do not always give all the details of what was included in repair costs. It is likely, for example, that car owners did not include a reserve in their reported costs. However, we can conclude that, from the late 1960s onwards, maintenance and repair costs increased at a much slower rate than the average salary, so automobile consumption generally became much more affordable. Still, repair costs were a significant cost item and, as we will see in the following section, many motorists preferred to repair their cars themselves, as they considered it less expensive.

DIY REPAIR AND AUTOMOBILE CONSUMPTION

As we have briefly discussed in previous sections, motorists started to repair their cars themselves long before the 1960s. Initially it was the lack of professional garages and other repair infrastructures that forced motorists without a chauffeur to self-repair. To acquire the necessary technical knowledge car drivers could consult a wide selection of manuals and self-help books. The well-known instruction book "Without a Chauffeur" was published in 13 editions between 1904 and 1930.[85] In addition, special repair guides offered systematic overviews of possible faults and their causes or guidelines on how to use one's senses to diagnose e.g. visible or audible symptoms.[86] Journals published similar self-help articles and advised motorists on useful tools and spare parts that should be taken on board for longer trips.[87] Hundreds of letters to the AAZ, a selection of which was regularly published in the "letterbox" column, testify to the numerous DIY repairs that were being carried out. It is interesting to note that the various repair manuals and readers' letters seldom mentioned economic reasons for self-repair. Motorists repaired their cars out of necessity or when they distrusted ad-hoc mechanics. The latter was often the case before car mechanics started to professionalise their trade in the late 1920s.[88]

---

84 Lotz, "Nach 40 Millionen km".
85 The first editions were entirely dedicated to motor cyclists.
86 See e.g. König, Adolf: Das Automobil und seine Behandlung, Berlin: R.C. Schmidt 1919; Heßler, Rudolf: Der Selbstfahrer, Leipzig: Hesse & Becker 1926.
87 See e.g. Schur, Rolf: "Werkzeugergänzung für Kraftwagen", in: AAZ 29, 44 (1928), p. 18.
88 The term "ad-hoc mechanic" has been coined by Kevin Borg to describe the numerous fitters, blacksmiths etc. that became car mechanics without special training at the

However, the bourgeois self-drivers of the interwar years perceived self-repair not only as a technical burden but also as a welcome means of social distinction. It enabled them to imitate the automobile consumption of the rich automobile pioneers, for whom mishaps and their elimination were an inherent part of the automobile adventure. Moreover, the new class of self-driving motorists were able to demonstrate their mastery of modern technology and prove their group affiliation by publicly self-repairing. Andrea Wetterauer interprets the demonstrated enthusiasm for and associated pride in self-repairing as an essential aspect of the bourgeois automobile culture of the 1920s.[89]

After the Second World War, it is noticeable that self-repair was initially somewhat absent in the automobile discourse in Germany: magazines published very few articles on the topic.[90] It was not until 1960 that self-repair regained wider attention in the automobile press. That year the motto "do-it-yourself" appeared for the first time in the *ADAC Motorwelt*,[91] and the following year, the new "do-it-yourself" column (*mach' es selbst*) was introduced, in which even demanding repair jobs such as checking wheel bearings were described.[92] In 1962, the first in a long series of repair manuals called "Now I help myself" (*Jetzt helfe ich mir selbst*) appeared on the market. They were written by the well-known automotive journalist Dieter Korp. Each manual was dedicated to a specific car model, and the series sold more than 2.5 million copies until 1975.[93] Another popular repair manual series, entitled "This is how you do it" (*So wird's gemacht*), published its first edition in 1966.

The huge success of self-help literature can partly be explained by the accelerating mass motorisation in West Germany from the 1950s onwards. The number of passenger cars went up from 518,000 in 1950 to 3.5 million in 1959, and

---

time when automobility was emerging: Borg, Auto Mechanics, ch. "Ad Hoc Mechanics". See also Krebs, "Sobbing, Whining, Rumbling"; id., "Notschrei".

89 Wetterauer, Andrea: Lust an der Distanz. Die Kunst der Autoreise in der "Frankfurter Zeitung", Tübingen: Tübinger Vereinigung für Volkskunde 2007; see also Schramm, "Konsumgeschichte".

90 This does not mean that motorists did not repair their cars themselves. It is relatively likely that during the late 1940s and early 1950s, when the supply of spare parts was still difficult, many cars were repaired by their owners. A few repair articles were published, see e.g. Anon.: "Was ist zu tun", in: ADAC Motorwelt 5, 2 (1952), p. 29; Brinzer: "Der Autofriedhof", in: ADAC Motorwelt 8 (1955), p. 335.

91 Anon.: "Wagenpflege leicht gemacht", in: ADAC Motorwelt 13 (1960), p. 272.

92 See e.g. JFD: "mach' es selbst", in: ADAC Motorwelt 16 (1963), p. 1168.

93 G. W.: "Basteln nach Handbüchern?", in: ADAC Motorwelt 28, 12 (1975), p. 19–21.

the total mileage from 5.49 million kilometres in 1953 to 50.3 million in 1959.[94] One important change in the interwar period was that more members of the lower middle class and working class started to own a car, and many of them first bought a second-hand car. In 1956, 44% of second-hand cars were bought by workers and employees, but they purchased only 30% of the new cars sold that year.[95] In 1961, more used than new cars were sold for the first time: 1.1 to 1 million.[96] Saving money was often mentioned as the prime reason for buying a second-hand car and carrying out repairs oneself.[97] After the oil crisis, journalists even diagnosed a "new austerity wave" that led to an increase in demand for second-hand spare parts.[98]

Another more general motive for self-repair can be found in the emerging DIY movement of the 1970s that encompassed many other craft activities.[99] The further professionalisation and commercialisation of DIY car repairs can be seen by the opening of garages that rented out lifts and tools and sold spare parts. The oil company Shell had still failed to establish a chain of DIY garages by the mid-1960s, but in the 1970s such garages could be found in many larger cities, and some had up to 70 work spaces.[100]

Saving money was not the only motive for motorists deciding to repair themselves. The self-help literature of the 1960s pointed out that self-repairing not only saved time and money; it was also "fun".[101] Car maintenance was referred

---

94   Anon.: Historische Anzahl an Kraftfahrzeugen und Personenkilometer nach Kfz-Typ in Deutschland in den Jahren 1906 bis 1959, https://de.statista.com/statistik/daten/studie/249900/umfrage/historische-entwicklung-von-kraftfahrzeugen-in-deutschland/#professional (accessed 22.02.2021).

95   König, Das Automobil, p. 127.

96   Anon.: "Die zweite Hand", in: Der Spiegel 22/1961, p. 24–33, here p. 24 and 25.

97   The book cover of the "Jetzt helfe ich mir selbst" book series stated "saves money and time". See e.g. Korp, Dieter: Jetzt helfe ich mir selbst: VW 1200, Stuttgart: Motor-Presse-Verlag 1965, front cover.

98   STR: "Immer mehr gefragt: Ersatzteile aus zweiter Hand", in: ADAC Motorwelt 28, 2 (1975), p. 4–8.

99   On the German DIY movement see Voges, Jonathan: "Selbst ist der Mann": Do-it-yourself und Heimwerken in der Bundesrepublik Deutschland, Göttingen: Wallstein 2017.

100   WM: "Selbstbedienung auch in der Werkstatt?", in: ADAC Motorwelt 25, 8 (1972), p. 59–63.

101   Korp, Dieter: Jetzt helfe ich mir selbst: VW Käfer 1200/1300/1500, Stuttgart: Motorbuch Verlag 1969.

to as a "hobby",[102] "a lively and gladly practiced 'sport'" that compensated for the lack of physical activity in white-collar jobs.[103] Another author suggested that "noble hobbyists, car enthusiasts and experienced amateur mechanics" would indulge in car repair "out of the pure desire for building and constructing [things]".[104] Self-help manuals also praised the general understanding of automobile technology as a value in itself; motorists would gain in self-confidence even if they did not plan to get their hands dirty themselves.[105] There were also plenty of male stereotypes served:[106] "The child in the man [must] play all his life. Among other things, with a used car!"[107] The *ADAC Motorwelt* stated that

"[n]o one works on his TV, no one repairs his oil heater, and hardly anyone mends his typewriter. Only with cars do even laymen tinker. And [they do so] with fervour and great joy and without fear of dirt and pinched fingers."[108]

Moreover, an advertisement for the *mot auto-journal* celebrated knowledgeable motorists as being more popular in their circle of friends.[109] Without any references to the DIY car repair practices of the interwar years, the DIY car repair of the 1970s articulated similar ideas that tinkering with car technology would foster male identities and help to build a distinct collective of knowledgeable motorists.

Market research and opinion polls show that these were not just vain hopes: in the 1969 DIVO survey, car repair ranked seventh in a list of the most popular hobbies. In 1971, an ESSO study found that car repair was a popular recreational activity: three out of ten motorists interviewed said that they repaired many

---

102 Anon.: So wird's gemacht: VW 1300 und 1200A, Bielefeld and Berlin: Delius, Klasing & Co. 1966, n. p. [preface].

103 Thaer, Albrecht/Korp, Dieter: Jetzt helfe ich mir selbst: Die Autokarosserie, Stuttgart: Motorbuch Verlag 1975, p. 10.

104 Landgraf, Jahn-Knut: So wird's gemacht: VW 1302/1302 S/1303/1303 S, Bielefeld/Berlin: Delius, Klasing & Co. 1973, n. p. [preface].

105 Korp, Dieter: Jetzt helfe ich mir selbst: NSU Prinz, Stuttgart: Motorbuch Verlag 1965, p. 9.

106 Anon.: So wird's gemacht: VW 1500/1600, Bielefeld/Berlin: Delius, Klasing & Co. 1969, n. p. [preface].

107 Lotze, Hedelore: "Das Auto und die Männer", in: ADAC Motorwelt 7 (1954), p. 711.

108 G. W., "Basteln".

109 mot auto-journal: advertisement, in: Korp, Dieter: Jetzt helfe ich mir selbst: Ford 12 M, Stuttgart: Motorbuch Verlag 1964, n. p.

things themselves. Moreover, 37% of car owners attested that car maintenance was a "pleasant" leisure activity. However, the survey also revealed that pleasure and economic necessity often went hand in hand because the wealthier motorists were, the less inclined they were to self-repair.[110]

Car mechanics perceived DIY repairers as unwanted commercial competitors. Already during the professionalisation of the trade in the 1930s, representatives of car mechanics guilds and manufacturers had argued that non-specialist craftsmen and laymen did not possess the necessary knowledge and skills to carry out repairs, and that self-repairs could even harm the operational safety of cars.[111] In the 1960s, the authors of self-help guides warned their readers to be careful during the warranty period because self-repairing could cause difficulties for warranty claims. They also emphasised that safety-related repairs, e.g. of brakes, should be left to professional car mechanics. Still, the guides explained how to carry out these repairs.[112] The representatives of car mechanics went one step further. In 1972, the Central Association of Car Mechanics warned: "the car becomes a weapon [...] when repairs are carried out by laymen, who are often unaware of how fatal mistakes can affect their own and others' safety."[113]

The association funded an advertising campaign with the slogan "do-it-yourself can be costly".[114] The German Association for Technical Inspections even lobbied for a ban on DIY car repairs. The well-known DIY magazine "Himself is the Man" (*Selbst ist der Mann*) countered this as a completely exaggerated criticism, but recognised at the same time that motorists should keep their fingers off the brakes and steering.[115]

CONCLUSION

To sum up, car repair was not just a means of maintaining technical functionality, and thus facilitating the consumption of (auto-)mobility; it was also an end in itself: repairing one's car fostered one's self-assurance and identity. The emer-

---

110 WM, "Selbstbedienung".

111 Krebs, "Notschrei".

112 See e.g. Korp, Dieter: "Das kann jeder selber machen", in: ADAC Motorwelt 21, 1 (1968), p. 38–43; Anon.: "Neue Bücher", in: ADAC Motorwelt 19, 9 (1966), p. 57.

113 WM, "Selbstbedienung", p. 62.

114 Das Kraftfahrzeug-Gewerbe: advertisement, in: ADAC Motorwelt 25 (1972), n. p.

115 Anon.: "Unser Kommentar", in: Selbst ist der Mann 1/1977, p. 11; see also Haberl, Fritz, "Das Kind nicht mit dem Bade ausschütten", in: Selbst ist der Mann 10/1977, p. 4.

gence and expansion of professional repair services facilitated the widespread purchase and use of automobiles during the interwar period and in particular from the late 1950s onwards. In addition, maintenance and self-repair were important moments in the individual and collective appropriation of the motor car. Furthermore, it is interesting to note that the bourgeois self-drivers of the interwar period and the working-class motorists of the 1970s voiced similar arguments for doing-it-yourself: mastering automotive technology was perceived as a way of increasing one's social status. At the same time, there were also significant differences as the DIY repairers of the 1970s were part of a larger DIY movement that reacted to the post-war reduction in working hours and included contradictory elements of progressive environmentalism and petit-bourgeois home cocooning.[116]

For the history of consumption, we can conclude that, until the 1950s, the cost of repairs was a decisive factor in determining who could afford to own a car at all. In addition, the initial unreliability of automobiles was an important argument not to buy a motor car. In other words, increasing the reliability of automotive technology was key for the successful establishment of the automobile consumer sector. This is also true for the emergence and expansion of a proper repair infrastructure that encompassed professional workshops and roadside assistance services. The field of car repair, including self-repair, was central to the emergence and development of automobility. It is likely that other consumer technologies, like radios and TV sets, required similar repair infrastructures, and the history of consumption still lacks a thorough investigation of the role of repair in consuming technology.[117] Likewise, the history of technology still needs to acknowledge the importance of repair for the diffusion of new consumer technologies. It would be worthwhile taking a second look at questions such as the development of the spare parts and accessories trade, gender roles in repair and self-repair, and the difficult socio-technical relationship between consumers and repairers.

---

116   Voges, Selbst ist der Mann.
117   See Weber, "Mending or Ending?".

Of Buses, Batteries and Breakdowns:

The Quest to Build a Reliable Electric Vehicle in the 1970s

Karsten Marhold

The 1970s were a window of opportunity to resurrect an old technology: the electric vehicle. A need for new business opportunities, saturated markets in countries where electricity was henceforth omnipresent, an emerging awareness of environmental issues, and two energy crises provided the environment to create renewed interest in electric road transport. In this chapter, I will discuss the activities of Électricité de France (EDF), the French state power company, and the German Gesellschaft für elektrischen Straßenverkehr (GES). The latter company was founded by the Rheinisch-Westfälisches Elektrizitätswerk (RWE AG), a major German electric utility, in order to develop electric vehicles on its behalf.[1] In both countries, the initiatives in the field came almost exclusively from the electricity industry – not from the car industry – and from these two companies.

In the first few sections I provide a short description of two electric vehicle test programmes carried out by EDF and GES, before discussing a series of problems discovered during these trials. I describe how the vehicles' batteries were identified by both companies in parallel as the most important and also problematic component of the vehicle. I then take a closer look at these problems, and argue that they were mostly related to maintenance, repair and reliability. These issues occupied the bulk of the engineers' attention, while also gener-

---

1   At EDF, the Research and Development Division (Études et Recherches) was in charge of the majority of the activities, but it often worked together with the Distribution Division (Direction de la Distribution), whose members were well placed to be test users for vehicles on the ground.

ating a series of connected problems related to the economic viability of the vehicles and the social context.

What I am describing here has traditionally been called a "reverse salient",
"critical problem" or "technological imbalance".[2] Melvin Kranzberg defines this
as "a situation in which an improvement in one machine" – for instance, modify-
ing cars or buses so that they can drive on electricity – "upsets the previous balance and necessitates an effort to right the balance by means of a new innovation".[3] In this case, it meant improving the battery. Similarly, Thomas P. Hughes,
who coined the term "reverse salient", describes a situation where an identifiable
part of a technical system lags behind the rest of its components and holds back
the further development and improvement of the whole, until the critical part itself is improved and allows the system to function correctly. This definition, as
we shall see, neatly fits the problems with batteries in 1970s electric vehicles.
They were a central critical issue within a larger technical system, with outsized
importance for the functioning of its other parts.

However, I also intend to show that this theoretical framework is not sufficient to capture innovation processes in all their complexity. In the conclusion,
therefore, I make some suggestions on how to integrate questions of use and
maintenance into the empirical analysis of 1970s electric vehicles and their batteries.

The sources I am using in this chapter are mostly drawn from the EDF and
RWE company archives. I remain close to the engineers' own perspective
throughout the chapter. This is because I am not trying to determine whether the
decisions they made were right or wrong, but rather to follow them through the
innovation process in order to understand "what they knew and how they knew
it".[4] However, I also present a battery maintenance process in more detail, in order to put the engineers' own claims into perspective.

---

2   Hughes, Thomas P.: Networks of Power. Electrification in Western Society, 1880–
    1930, Baltimore: Johns Hopkins University Press 1983, p. 79–105; Kranzberg, Mel-
    vin: "Technology and History: 'Kranzberg's Laws'", in: Technology and Culture 27
    (1986), p. 544–60, here p. 549. See also Lee Vinsel's comment on Kranzberg's sec-
    ond law from a maintenance-oriented perspective, "Kranzberg's First and Second
    Laws – Technology's Stories", https://www.technologystories.org/first-and-second-
    laws/ (accessed 04.07.2019).
3   Kranzberg, "Kranzberg's Laws", p. 549.
4   I am alluding to Vincenti, Walter G.: What Engineers Know and How They Know it:
    Analytical Studies from Aeronautical History, Baltimore: Johns Hopkins University
    Press 1997.

TESTING ELECTRIC UTILITY VANS AND BUSES: THE SETUP

The sourcing and construction of the vehicles and parts in question, as well as the relationships of EDF and GES with the automotive industry, deserve a discussion of their own. The vehicles discussed below, the Renault 4 delivery van and the MAN SL-E bus, were based on internal-combustion engine (ICE) series models converted to electric drive.[5] Despite the fact that purpose-design vehicles existed and were built and tested by both companies, such conversion designs were the norm, especially in the early years. Although Renault and MAN built the vehicles and delivered them to the electric utilities, they did so in close cooperation with GES and EDF and according to their specifications, and in both cases the vehicles were the result of a cooperative effort between a number of companies that was coordinated by electric utilities. This included the vehicle industry as suppliers of the chassis, body and conventional vehicle parts, battery manufacturers such as VARTA AG and Fulmen, as well as suppliers of electrical components and motors (BOSCH AG). Indeed, both GES and EDF were aware that they would have neither the ability nor the interest to produce vehicles themselves in the short or long term. In an early phase, however, they accepted the need to support efforts to develop electric vehicles with specialised know-how and resources.[6]

---

5   For reasons of space and precision, in this chapter I concentrate on these two particular vehicle types. It must be noted, however, that both utilities developed and tested delivery vans and buses and also ventured into the realm of private vehicles, in particular GES towards the end of the decade. As far as the timeframe goes, most of the activities in this area began as early as the mid-1960s and declined from the mid-1980s onwards, without ever coming to a complete halt. Despite the deliberate limitation in this paper, the selection of vehicles is therefore representative of both utilities' electric vehicle programmes.

6   For introductions to the EV projects, see Döring, Peter/Thomas, Hans-Georg: "Vom 'Pfennigspaß' zum Milliardengrab? RWE und die Entwicklung eines Elektroautos in den Jahren von 1964 bis 1986", in: Horstmann, Theo/Döring, Peter (eds.): Zeiten der Elektromobilität. Beiträge zur Geschichte des elektrischen Automobils: Beiträge der Tagung des VDE-Ausschusses "Geschichte der Elektrotechnik" in Kooperation mit dem VDE Rhein-Ruhr e.V. vom 7. und 8. Oktober 2010 in Dortmund, Berlin: VDE 2018, p. 123–180; Griset, Pascal/Larroque, Dominique: L'odyssée du transport électrique, Paris: Cllomedia 2006; Nicolon, Alexandre: Le vehicule électrique: mythe ou réalité, Paris: Éditions de la Maison des Sciences de l'Homme 1984; Callon, Michel: "L'État face à l'innovation technique: le cas du véhicule électrique", in: Revue fran-

EDF's testing programme began in 1973 and initially concerned 36 Renault 4 (R4) vehicles, a fleet that increased to 90 vehicles two years later.[7] Testing consisted of two stages: in stage one, cars were delivered to EDF's own distribution centres in the Greater Paris region, where employees would use them to complete their daily work schedules – performing regular maintenance on the distribution infrastructure – and report back to EDF. After six months of this preliminary test, EDF extended its trial for another six months to private and public partner companies in the Paris region who would drive the R4s, on the condition that they could provide fixed routes in advance that corresponded to the range of the vehicles.[8] The partners included aircraft engine manufacturer Snecma in Melun, local post offices, waterworks and the Melun prefecture. In both cases vehicles were driven under everyday conditions and on public roads, although EDF tried to make sure that there were clearly defined boundaries: the vans were used for mail distribution, client visits or simply to run errands during the day or drive to the canteen. The cars used in these tests managed average ranges of between 20 and 30 km per day, or between 250 and 500 km per month, depending on the location and specific use. By 1974, the EDF vehicles had been driven for about 120,000 km in these two tests.[9] In parallel, EDF continued testing a number of R4s and a variety of other types of vehicles at its research centre in Les Renardières close to Paris, and did so for several years after the on-road tests. By 1979, those vehicles had been driven for more than a million kilometres in total.[10]

GES's bus testing programme began in autumn 1974, after a preliminary phase during which two prototypes had already been put into service for a few months in Koblenz and other German cities.[11] When the on-road trials began in

---

çaise de science politique 29 (1979), p. 426–47; Callon, Michel: "The Sociology of an Actor-Network: The Case of the Electric Vehicle", in: Callon, Michel/Law, John/Rip, Arie (eds.): Mapping the Dynamics of Science and Technology, London: Palgrave Macmillan 1986, p. 19–34.

7   Anon.: "Véhicules électriques. Expérimentation et proposition de programme 1975", 060134, D0000252716, Archives EDF.

8   Heurtin, J.: "Expérimentation dans les Villes Nouvelles : Examen des réponses au questionnaire 'véhicule électrique'", 9 May 1974, 823334, Archives EDF.

9   EDF, Direction des Études et Recherches, département Applications de l'Électricité: Leaflet "Véhicule électrique", 1974, 823349, 29105, Archives EDF.

10 Pasquini, P.: "Bilan 1979 de l'expérimentation des véhicules électriques aux Renardières", 24 Mar. 1980, B0000428243, D0000259426, Archives EDF.

11 "'Elektro-Bus', Referat H. Dir. Scheffel (KEVAG) am 27.1.1971 anläßlich der Vorstellung in Koblenz", 6153, Historisches Konzernarchiv RWE.

earnest, twenty MAN buses, also a conversion design from a regular series model, were put into service in two cities in North Rhine-Westphalia: seven buses on one line in the city of Mönchengladbach, followed a few months later by thirteen buses on two lines in Düsseldorf. The Mönchengladbach test ran until 1981, when public funding came to an end, while the Düsseldorf trial continued until 1987. Each bus covered around 300 km per day when in service. Taking the two cities together, this added up to 225,000 km per month, and over 5 million km by 1981. The distance covered was therefore by several magnitudes greater than in the EDF trials – but also concerned an entirely different class of vehicle.

The enormous difference in daily range compared to EDF's vehicles was made possible by a system of battery-changing stations developed by GES that allowed buses to replace a depleted battery with a charged one in a matter of minutes. Unlike EDF's delivery vans, which had permanently installed batteries, the GES buses would change theirs 5 to 9 times per day. In heavier traffic, more changes were needed to reach the daily range of about 300 km.[12] To make the changing process easier, the batteries were not integrated into the vehicle body but rather housed in a trailer attached to the bus. To change the battery, a bus driver would drive the bus in front of the station, which resembled a container and had an opening through which batteries could slide in and out. Either the driver or a specialised mechanic would then swap the batteries using a fully automated process that could be controlled from a remote control panel located outside the station. In the station, batteries were stored on a rack with threaded rods, which allowed them to be moved upwards and downwards depending on whether they were ready to be exchanged or had to be stored. Towards the end of the testing phase, in 1981, GES abandoned this system in order to replace it with one that allowed the buses to be recharged at each stop. It should be noted that no such system, of either kind, was ever used by EDF.

MAINTENANCE, REPAIR AND RELIABILITY: PROBLEM AREAS

The result of the tests was twofold: on the face of it, the vehicles of both EDF and GES were proven to be functional. It was demonstrated that they could, in principle, meet the conditions that had been set for the trials. Engineers and

---

12 Döring/Thomas, "Vom Pfennigspaß zum Milliardengrab", p. 144; Moneuse, M.: "Mission en Allemagne Fédérale des 3 et 4 Novembre 1975. Visite d'installations de la GES à Essen et Düsseldorf", B0000428238, D0000259410, Archives EDF.

managers from both companies repeatedly pointed this out in various reports on the tests. They did realise, however, that the range, speed and weight of the vehicles were grossly inferior to conventional vehicles. For instance, a survey among EDF's test users revealed that most of them considered the range of the vehicles to be "insufficient" or "totally insufficient", but none complained that this had prevented them from using the vehicles as planned during the trial. The head of GES acknowledged after several years of testing that electric vehicles had the "uncontested drawback of a limited operating range".[13] When French carmaker Renault decided to reduce its efforts to develop electric vehicles in 1976, the company based its decision on such performance criteria. For Renault, an electric car's properties could be converted into an equivalent internal-combustion engine car with a maximum speed of just 70 km/h, a tank with a capacity of 5 litres that needed 8 hours to be refilled, and a vehicle that was constantly at its maximum weight and able to carry only half the load.[14] Similarly, a manager of the German car manufacturer Daimler pointed out that the battery block of an electric vehicle was 120 to 170 times heavier than the equivalent in petrol, while coming at a higher cost.[15]

Despite these obvious shortcomings of electric vans and buses in terms of raw performance, which became obvious in the trials, EDF and GES remained optimistic that electric vehicles could find their niche on the market if they could at least be made more reliable, easier to repair and low maintenance.[16] If this was possible, then maybe they could be used in ways similar to the trials: in a context where delivery vans were needed for predictable, clearly-defined and short trips, or in cities that might be willing to invest in electric buses in order to resolve problems with noise and reduce air pollution. When they presented their insights from the tests at the 1976 Electric Vehicle Symposium, an international conference on the matter, engineers from EDF and from GES independently concluded

---

13 Müller, Hans-Georg: "Energiewirtschaftliche Überlegungen zum elektrischen Strassenfahrzeug", in: ZEV-Glasers Annalen 103, May 1979, p. 233–236, Historisches Konzernarchiv RWE; Heurtin, Expérimentation dans les Villes Nouvelles.

14 TREGIE: Véhicules électriques. Position Renault, March 1976, 823338, 29094, Archives EDF.

15 Breitschwerdt, Werner: "Letter to Helmut Meysenburg", 8 Jul. 1980, 6155, Historisches Konzernarchiv RWE.

16 Hagen, H./Zelinka, J.: "The MAN Electrobus. Experience Gained in Large-Scale Tests", 1976, AVERE Archives; Heurtin, J./Moneuse, M.: "Expérimentation de véhicules électriques légers et lourds à EDF. Exposé des problèmes qui se posent", 1976, AVERE Archives.

that, instead of performance, it was reliability that should become front and centre in their further work on electric vehicles.

Indeed, in addition to the mediocre speed, range and weight, both companies' vehicles were equally unfavourable in terms of reliability when compared with conventional vehicles. The problems began with vehicle parts that had not been modified from the conventional versions of the vehicles. They suddenly failed or needed to be modified when used in an electric vehicle. An example of this is tyre wear: due to the higher weight of the GES buses, the service life of tyres on the drive axle was reduced to 20,000 km instead of a standard 80,000 km.[17] Added to this were overheating and jamming pressured-air systems in buses; unusually high axle wear due to the higher weight of the vehicles; blown fuses; or buses starting to move unexpectedly because of faulty ignitions.[18] Such problems were irritating, constant reminders to engineers that developing electric vehicles did not mean just replacing vehicles' engines but that they were dealing with complex systems that had to be adjusted to real-world operating conditions in various ways. While the problems with conventional vehicle parts could eventually be brought under control, they were overshadowed by vastly more important issues with batteries, the central component of the new vehicles.

BATTERIES: THE KEY COMPONENT

It had been expected from the outset that batteries would be a delicate issue, and tests confirmed that they were indeed the most crucial and problematic component of the vehicles. Engineers at both companies framed the problem above all in terms of battery lifetime. In the case of EDF, the reason for this was that the first-generation batteries used in 1973 lasted only 50 cycles on average, and because of this limitation were quickly understood to be unpractical for any real-life use.[19] A second generation of batteries with lower energy density but higher weight, subsequently installed in the vehicles, was reported to have reached 200

---

17 GES: "Ergänzung zum Besprechungsvermerk vom 30.3.1976 über den Einsatz der Elektrobusse bei den Stadtwerken Mönchengladbach", 6 Apr. 1976, 11142, Historisches Konzernarchiv RWE; see also Döring/Thomas, "Vom Pfennigspaß zum Milliardengrab", p. 145.

18 GES: "Zwischenbericht vom 31.12.74", 11131, Historisches Konzernarchiv RWE.

19 A "cycle" refers to a charge-discharge operation. MIT Electric Vehicle Team, "A Guide to Understanding Battery Specifications", Dec. 2008, p. 3.

cycles in at least one instance; and a third generation of batteries used from 1976 onwards doubled this value again to 400 cycles. EDF's engineers therefore concluded from the first tests that lifetime was the "essential problem" to be resolved before turning to performance again.[20]

Battery lifetime was also a concern for the GES buses. Although GES originally claimed in 1971 that battery lifetime was approaching 100,000 km,[21] the results of the first bus tests were sobering: first-generation batteries began to fail "prematurely" in 1976 and unexpectedly reduced the availability of the buses.[22] Nevertheless, the bus batteries were able to reach a lifetime of about 1,000 cycles.[23] In kilometres served, the first generation of batteries only lasted for about 47,000 km, whereas the subsequent generation managed up to 65,000 km – still far short of the initially hoped-for 100,000 km.[24] Nevertheless, when GES engineers reported on the bus trials, they did not point first and foremost toward to the life cycle of the batteries. Changing stations eliminated the risk that a dead battery would immobilise a bus for too long. Yet batteries "dying" was by no means the only possible problem: "increased consumption of water at high operating temperature, more pronounced pollution in road traffic, more frequent checking caused by five cycles per day, extremely high proportion of peripheral apparatus, [and] requirements higher than usual for the insulation value of the batteries" were problems only discovered during the first on-road tests.[25] Many of these were rather unexpected and would only appear once the vehicles were driven in real-world conditions. Better insulation of the batteries was needed for instance because in winter, on icy roads, thawing salt got into the battery blocks and compromised insulation values.

---

20 Heurtin/Moneuse, Expérimentation de véhicules électriques légers et lourds.

21 GES: "Gedanken von RWE, GES und SELAK zum elektrischen Straßenverkehr", 6153, Historisches Konzernarchiv RWE.

22 GES: "Zwischenbericht vom 11.9.1976", Zwischenberichte an das Ministerium für Arbeit, Gesundheit und Soziales, 11132, Historisches Konzernarchiv RWE.

23 Indeed, there is a considerable difference between this figure and the life cycle of the batteries that EDF used. This can be explained by two factors: first, the bus batteries were being charged in a controlled environment in the charging stations, which certainly contributed to their longer overall lifetime. Sources are also unclear as to whether the same standards were used on both sides to determine whether a battery was "dead".

24 GES: "Zwischenbericht vom 6.10.1980", Zwischenberichte an das Ministerium für Arbeit, Gesundheit und Soziales, 11133, Historisches Konzernarchiv RWE.

25 Hagen/Zelinka, The MAN Electrobus.

As a consequence, GES noted between 13 and 18 such incidents with batteries per 10,000 km, or in other words, a problem with a battery about every 500 km during the first year of operation of the MAN electric buses in Düsseldorf and Mönchengladbach. When one adds other issues with conventional vehicle parts and changing stations, this meant that buses were taken out of service by some kind of incident about every other day on average. That was a rate nine times higher than that of conventional buses.[26] During the initial period, the buses were available only 38% of the time, and had to be replaced by conventional buses for the rest.[27] Therefore, even though the life cycle of individual batteries was less of a concern due to the changing system, reliability, maintenance and troubleshooting were still pressing concerns for GES as well.

ECONOMICS: THE COST OF MAINTENANCE

One of the major concerns for EDF and GES was to make the economic case for electric vehicles and compare them favourably to conventional vehicles in terms of costs. This led to the first "well-to-wheel" calculations, which compared the overall energy efficiency and costs of conventional and electric vehicles. In other words, such analyses did not just take into account fuel and electricity costs needed to drive the vehicles, but also the efficiency of their motors and how much primary energy they consumed overall.[28] The patchy reliability record of both EDF's and GES's vehicles also had important implications in this respect, however. Maintenance requirements were so high that they played a major part in making the vehicles more expensive and less economically viable.

For EDF, the limited lifetime of the batteries it tested was above all a problem in economic terms. EDF engineers duly noted that batteries were already the most expensive part of the vehicle, and therefore would have to be changed as rarely as possible. The *technical* operation of changing batteries was not out of the ordinary in terms of difficulty, and could even be sped up and automated, as

---

26 Döring/Thomas, "Vom Pfennigspaß zum Milliardengrab", p. 145.
27 GES: "Niveau de développement des bus électriques desservant une ligne fixe et exploitation de ces bus à Mönchengladbach et Düsseldorf-Benrath (situation au 30 juin 1977)", 1977, B0000428237, D0000259406, Archives EDF.
28 Müller, Hans-Georg: "Elektrischer Strom ist für den Nahverkehr auf der Straße eine kurzfristig verfügbare, bisher zu wenig beachtete Alternative zu Kraftstoffen aus Erdöl", Vortrag auf der Mitgliederversammlung der Deutschen Gesellschaft für elektrische Straßenfahrzeuge e.V. – DGES – am 9. Mai 1980 in Berlin, May 9, 1980, Historisches Konzernarchiv RWE.

GES's battery changing stations demonstrated at the same time across the border in Germany. With hypothetical, cheap "throw-away" batteries, limited lifetime might have been a lesser problem. The reality, however, was quite the opposite. Even with batteries reaching 400 charging cycles, as EDF managed to do by the mid-1970s, it was estimated that they would represent 20–25% of the operating costs of the vehicle. EDF was forced to conclude that such costs were "prohibitive". Moreover, real-world use of the vehicles that led to inconsistent charging patterns penalised the vehicles economically by reducing the lifetime of the batteries and therefore increasing operating costs.[29]

For GES, too, maintenance and breakdowns were above all a problem because of high cost. The bus batteries were more reliable than those of EDF's utility vehicles, but maintenance remained an important issue because the costs of upkeep pushed GES's budget to the limit. The company's contracts with local public transport operators stipulated that the latter would assume operating and maintenance costs up to the value of the equivalent needed for diesel buses. As breakdowns were much more frequent than for conventional buses, however, the operators quickly made it clear that GES would have to cover the excess maintenance costs.[30] Such a demand strained GES's budget, which was almost entirely dependent on funding from parent company RWE, and also threatened to drain resources that could otherwise have been used for development work and engineering. Moreover, for GES the problematic cost of maintenance was not the basic price of vehicle parts, but rather the cost of personnel and labour needed for maintenance. When reporting on the bus trials, a GES engineer noted that not only had the incidents "far exceeded" what had initially been thought, but also that this was mostly because specialised staff were needed to carry out maintenance and troubleshooting, and thus had to be hired in sufficient numbers. This was one more reason, therefore, to prioritise reliability when further improving the vehicles.[31]

Moreover, the battery-changing system did not help in this regard, insofar as it required GES to hold several batteries per vehicle in reserve. Not only were the batteries expensive in themselves, but the greater the number available, the

---

29 Heurtin/Moneuse, Expérimentation de véhicules légers et lourds; see also part 4 below.

30 GES, Ergänzung zum Besprechungsvermerk vom 30.3.1976 über den Einsatz der Elektrobusse bei den Stadtwerken Mönchengladbach, April 1976, 11142, Historisches Konzernarchiv RWE.

31 GES: "Zwischenbericht vom 17.4.1975", Zwischenberichte an das Ministerium für Arbeit, Gesundheit und Soziales, 11131, Historisches Konzernarchiv RWE.

higher the maintenance expenses. In the end, GES decided to abandon the changing stations for these reasons, and replace them with a catenary-based system that allowed short, intermediate charges at certain stops on the bus routes. Complete charges were done when buses entered the depot, usually overnight. This system ensured that from that point on, only one battery per bus was needed.[32]

CASE STUDY: A BATTERY MAINTENANCE MANUAL

To give an impression of how tedious maintenance procedures were and how carefully they had to be carried out, I will discuss a battery maintenance manual for a VW delivery van that was shared by GES with colleagues at EDF during a vehicle exchange between both companies. EDF and GES had decided to work together and sign a cooperation agreement in 1974, and as part of this collaboration, GES sent the VW vehicle to France to be tested by EDF in the Renardières research centre, while EDF sent a Renault 5 – the successor to the R4 described in this paper – to Düsseldorf in return. The manual was thus shared between the electrical engineers at both companies. This suggests that, even with the requisite knowledge and education that they possessed, battery maintenance was not a trivial process. Furthermore, the manual is revealing in that it can be used to calculate the time needed to maintain a battery, a crucial factor especially for fleet vehicles that are expected to be available as frequently as possible. Finally, by looking at the procedure in detail, it is possible to qualify the claims made by engineers about maintenance being complicated and labour-intensive.

The maintenance procedure described in the manual had to be carried out at least every 15 (+/-1) charge/discharge cycles of the battery. Assuming one charging cycle per day, this implies that the battery had to undergo maintenance every two weeks. In total, the manual contains 16 steps, eight of which refer to measurements and note-taking while the rest concern actually handling, cleaning and charging the battery.[33] At the beginning of the maintenance process, after some initial measurements (steps 1–4), the battery first had to be fully charged (5). This took 3 to 8 hours, depending on the type of charging equipment and charging procedure being used. This was followed by insulation measurements (6). The values were to be measured manually using a multimeter or an insula-

---

32 Döring/Thomas, "Vom Pfennigspaß zum Milliardengrab", p. 146.
33 GES: "Dossier d'information sur les batteries de traction. Communication technique N°18. Entretien des batteries de traction", 1975, 060110, D0000252644, Archives EDF.

tion monitoring device, and then calculated using a specified formula depending on the equipment used. Individual battery cells then had to be topped up with distilled water (7) according to specifications.[34] The next steps (8–10) had to do with cleaning the battery and tray: first, the maintainer had to check whether there was water in the battery tray and remove it if needed. The maintainer then had to clean dust and oxidation off cables and contacts and grease them to prevent them from getting dirty. If the previously measured insulation value was below a certain threshold, the entire battery had to be washed carefully, making sure non-distilled water would not enter the individual battery elements. Once cleaned, meters and indicators had to be checked to make sure they functioned correctly (11). In the next step the battery had to undergo an equalising charge, in other words a slow charge in order to ensure that all cells were charged as equally as possible (12). The manual states that with a standard electric outlet, this charge would take a minimum of 24 hours. Contrary to the initial charge, where an optimised charging curve was possible, no faster options are mentioned. The equalising charge was considered successful once acid density between cells and voltage remained constant for two hours; if not, the process had to be started again after a one-hour break. Final measurements and note-taking were then required (13–15). The last step (16) of the manual specifies how batteries held in storage were to be handled.

Given the large number of steps, different areas to pay attention to, specialised measuring equipment needed, calculations to be made and overall knowledge required, it is sensible to conclude that this procedure could not be carried out by just anyone driving the cars, but that it required skilled personnel. To borrow an image from Gijs Mom, the batteries needed "the constant attention of a physician and a trained nurse".[35] Moreover, correct maintenance and charging were crucially important in order to guarantee the maximum life cycle of the batteries. It is therefore understandable that it was an important cost factor to hire the staff needed, which in turn served as a guarantee that the procedures would be carried out as diligently as possible. More importantly, however, the process was very long and had to be carried out very frequently. As mentioned above, this was especially a problem for vehicles used in fleets. Using the times

---

34 As is standard for high-capacity batteries, the lead-acid traction batteries used in the VW Transporter were composed of a number of individual cells arranged in a tray and connected to each other.

35 Mom, Gijs: The Electric Vehicle: Technology and Expectations in the Automobile Age, Baltimore: Johns Hopkins University Press 2013, p. 287.

given in the manual, it appears that under ideal conditions, the charging time alone would be 27 hours. If only a standard outlet was available, the initial charge time would increase to 8 hours, and it can be assumed that equalising charges took longer in most cases than the minimum duration of 24 hours. In this more realistic scenario, charging time during maintenance was at least 32 hours. To this had to be added the time needed for topping up with water, taking measurements and cleaning the tray, for which no precise durations are specified in the manual but which can be estimated to have taken several additional hours. In total, a safe estimate would be that the maintenance process left a battery out of service for at least two full working days every other week, tying up specialised maintenance personnel. If the battery was fixed in the vehicle and could not be removed, this would mean that the vehicle was unavailable during this time. If, as in the case of GES while the changing stations were in use, the battery was removable, another one had to be made available as a replacement during the maintenance process.

PEOPLE: ELECTRIC VEHICLES IN USE

Maintenance was thus complex, very frequent and time intensive, and batteries had to be handled with extreme care and attention. Not only did these requirements drive up costs, but drivers and mechanics with sufficient knowledge were constantly in short supply. Drivers of EDF utility vans suffered from "range anxiety"; they often used the vehicles only for very short trips, and only if they could be sure that the range of the vehicle was sufficient. As a consequence, batteries were constantly discharged by only half their capacity, and in some of the tests by just 30%, which had not been expected initially. Such irregular use of the vehicles, with partial discharges and recharges, made the charging process less efficient and could reduce lifetime further. Moreover, such a pattern of use made it difficult for EDF to draw conclusions about the maximum range to be expected from the cars, as almost no drivers ever attempted to fully exhaust their vehicles' batteries.[36]

Similar issues arose with GES buses. At the beginning, matters were again complicated by the existence of the battery changing stations. It had initially been planned that specialised personnel would carry out the changing procedure. Later on, as part of the effort to drive down costs, GES considered having driv-

---

36 Heurtin, J.: "Compte-rendu de la réunion du groupe de travail 'véhicule électrique' du 26 septembre 1973", 823334, Archives EDF.

ers perform the changes themselves. This led to questions about training and the operating security of the stations.[37] The problems that hindered this transition ranged from the mundane to the dramatic. GES struggled to convince the bus drivers' professional society to agree to its proposed measures to protect drivers needing to change a battery when it rained. GES's proposal to equip drivers with a "jacket and umbrella" was roundly rejected as "unacceptable", and the association demanded that either the station should be operable remotely from inside the bus or a roof should be installed.[38] At the other extreme, in February 1980, a mechanic operating a station suffered a fatal accident.[39] As a consequence, the professional association concerned suspended the operation of changing stations pending the installation of better security systems.[40] In the meantime, however, GES had decided to abandon the system altogether.

The problems did not end there. Indeed, as the buses now had fixed batteries, the drivers and regular vehicle mechanics had a greater responsibility to correctly charge them, especially at the end of the day when buses entered the depot. As a consequence, GES found that this led to new problems precisely because drivers were not correctly following procedures when buses were in the depot. Sometimes they were disconnected from chargers at a fixed hour, although they were not supposed to leave the depot until later on; they were not reconnected to chargers when they were not going out that day and used only as reserve vehicles; they were not charged immediately after entering the depot at the end of service; when they were entering the depot in need of repair, they were put on the repair track instead of being connected to a charger; and so on. As a consequence, and initially unbeknownst to GES, buses had to be towed several times with depleted batteries, although the catenary and intermediate-charging system was designed to make sure this would never happen. The problem was aggravated by frequent changes in personnel at the depots, which made it difficult to

37 GES: "Besprechungsvermerk", 25 Mar. 1976, 11142, Historisches Konzernarchiv RWE.

38 GES: "Besprechungsvermerk, Einsatz der MAN-Busse in Mönchengladbach, Auflagen der Berufsgenossenschaft zur Bedienung der Wechselstation durch die Fahrer", 3 May 1976, 11142, Historisches Konzernarchiv RWE.

39 GES: "Zwischenbericht vom 2.9.1980", Zwischenberichte an das Ministerium für Arbeit, Gesundheit und Soziales, 11133, Historisches Konzernarchiv RWE.

40 Berufsgenossenschaft der Feinmechanik und Elektrotechnik: "Schreiben an GES mit Besichtigungsbericht vom 27.2.1980", 29 Feb. 1980, 11142, Historisches Konzernarchiv RWE.

maintain a level of training sufficient for correct operation.[41] Precisely because resident vehicle mechanics lacked the specialised knowledge needed to maintain electrical components and batteries, GES reluctantly had to resort to carrying out the maintenance itself or use other electrical engineering companies for such tasks, so the related expenses remained high.

REMEDIES: GETTING TO GRIPS WITH BATTERIES

Once maintenance and reliability had been identified as being of paramount importance, both EDF and GES began to focus their research and development efforts on these areas. For GES, "incident reduction" became a mantra for the remainder of the bus tests. A task force named "Basic Battery Issues", headed by GES, brought together representatives of battery manufacturer VARTA, W. Hagen, Volkswagen and GES to identify and develop solutions to known battery problems. The group identified six problem areas and underlined the complexity of the issue. Solutions were needed in the areas of "charging, monitoring, construction, climatisation and tray, discharging, after-treatment and regeneration".[42] Apart from this research group, GES implemented solutions on the ground in cooperation with battery supplier VARTA: later generations of batteries were equipped with automatic water replenishment systems and acid circulating pumps, eliminating several steps from maintenance procedures. The tweaks were largely successful: after seven years of operation, incident rates hovered around 120 over a period of three months, down from close to 500 at the beginning. By the end of the trial, the buses were available 96% of the time, similar to what could be expected from conventional buses.[43]

EDF's response to problems with inefficient and partial charging cycles was a testing programme for batteries in cooperation with GES in addition to the on-road tests. Having signed a mutual cooperation agreement, the first step for the two partners was to decide on a standardised testing procedure including different charge-discharge routines. This protocol was then applied to the different

---

41 GES: "Vermerk T-202: Einsatz der MAN-Elektrobusse in Düsseldorf-Benrath", 11141, Historisches Konzernarchiv RWE.

42 "Besprechung des AK 'Grundsatzfragen der Batterien', 9–10.5.1983", 11135, Historisches Konzernarchiv RWE.

43 Döring/Thomas, "Vom Pfennigspaß zum Milliardengrab", p. 145; GES: "Anlage 1 zum Schreiben der GES mbh, Düsseldorf, vom 22.6.77 an Herrn Dir. Dr. G. Klätte, RWE-Vorstand", 1977, 6154, Historisches Konzernarchiv RWE.

batteries the companies were using in their vehicles. As a result, it turned out that it was perfectly possible to achieve more than 1,000 cycles with a battery under laboratory conditions. However, EDF's experience on the ground was confirmed by the fact that the charging patterns needed to achieve such a high life cycle were less than ideal for real-life use: the best results were obtained with batteries that were charged and discharged from 75% to 25% and 50% to 0%. As we have seen, however, drivers would typically discharge their batteries from 100% to 50%. Using such a pattern, even by the end of the 1970s, the best result achieved was 300 cycles.[44]

ANALYSIS: REVERSE SALIENTS, MAINTENANCE AND TECHNOLOGIES-IN-USE

As mentioned in the introduction, it is certain that if one had to identify a reverse salient holding back the development of electric vehicles tested by EDF and GES in the 1970s, it was their batteries. However, it appears that simply identifying the battery as *the reverse salient* does not do justice to the complex problems that arose and that I have described above.

The first thing to note is that batteries were highly complex systems in themselves, as the variety of battery "problem areas" identified by GES's task force underlines. Summarising its work, one engineer noted that all problem areas mutually influenced each other and that it was essential to regard batteries as a system in themselves. But more importantly, describing batteries as the reverse salient does not help us to understand how exactly the batteries had to be improved, or what criteria were used to determine when their performance was finally sufficient. What were the engineer's goals? Certainly, they were aware that batteries were the most problematic component, or subsystem, of the vehicles. But as we have seen, they could have tried to improve performance first – high speeds, long range, large carrying capacity – or decide to focus on maintenance and reliability. Why did they choose the latter? Why resolve the "technological imbalance" in this way and not another? What does it mean, in fact, to bring a reverse salient "in line" with the rest of the system, or to "balance" it?

To shed light on these questions, I believe it is necessary to employ some concepts from more recent literature in the history of technology and apply them

---

44 "Note d'information 79-02", 5 Apr. 1979, B0000428237, D0000259406, Archives EDF.

to the present case: first, "technology-in-use", and second, ideas specifically focused on maintenance.

David Edgerton in particular has pointed out that historians of technology have mostly focused on invention processes, and have had a tendency to neglect the use of artefacts. In other words, historians have not paid enough attention to how things are actually being used, regardless of what engineers or inventors might have had in mind when designing them or testing them under lab conditions.[45] EDF and GES engineers themselves did not see it this way: testing the vehicles, using them and subjecting them to real-world conditions was crucial in revealing the battery as the principal reverse salient. At both GES and EDF, engineers were happy with the decision to begin testing the vehicles sooner rather than later, which suggests that they were very sensitive to the importance of actually using the vehicles in order to determine further development steps. When one GES engineer reflected on the bus testing programme, he self-critically asked the question of whether it might not have been premature to put the buses out on the roads, given the high number of incidents. But his answer was unequivocal: to prove in principle that one could build an electric bus, a prototype would have been sufficient, but to "lay the groundwork" for electric road transport as a system, there was no way around testing the buses in regular service.[46] The same can be said about EDF's delivery vans. Not only were they driven on public roads; they were ultimately driven by partner companies over which EDF had no direct control, proof that getting as close as possible to real-world use conditions was crucially important for the French engineers, too. In other words, engineers understood that the final form of technological artefacts follows failure as much as it follows function.[47] For failure to occur, however, things have to be given the chance to fail, and therefore the vans and buses had to be actually driven.

The early emphasis on use and the decision to submit the vehicles to real-world conditions in turn revealed the central problem that I have discussed in this chapter: that the biggest challenges engineers faced were to be found in the areas of maintenance and reliability. Recently, a growing number of STS scholars have pointed out that invention and innovation have in the past too readily

---

45 Edgerton, David: "De l'innovation aux usages. Dix thèses éclectiques sur l'histoire des techniques", in: Annales. Histoire, Sciences Sociales 53 (1998), p. 815–837.

46 GES: "Zwischenbericht vom 14.5.1976", Zwischenberichte an das Ministerium für Arbeit, Gesundheit und Soziales, 11131, Historisches Konzernarchiv RWE.

47 "Form follows Failure" as a design principle is discussed in Petroski, Henry: The Evolution of Useful Things, New York: Knopf 2010, p. 22–33.

been identified with the creation of new and shiny things, as well as the person-alities of ingenious inventors and "innovators".[48] But more mundane activities such as repair and maintenance, "tweaking", "hacking" and gradual improve-ments matter at least as much as making ground-breaking inventions. Engineer-ing, as Walter Vincenti has remarked, is first and foremost a problem-solving ac-tivity.[49] The history of EDF's and GES's electric vehicles provides good evi-dence for this type of argument. After all, no fundamentally new components or artefacts were involved in the process, and most of the work indeed focused on solving problems that were discovered during the tests, in order to gradually eliminate them. The end result, of course, was improved, and therefore "new", buses and delivery vans.

Recent literature on maintenance, however, tends to distinguish between in-novation and maintenance as separate or even contradictory processes. In other words, it gives the impression that maintenance only begins to matter after a thing has been invented, has been used and has broken down.[50] But it is not only after innovation that maintenance begins to matter. Rather, maintenance is part and parcel of all stages of the lifetime of an artefact. Without question, electric vehicles were considered innovative in the 1970s. But the point I would like to make is that already during the innovation process, maintenance and repair were crucially important. In fact, they were even identified as the principal problems to be resolved and were considered to be "life-threatening" issues for the success or failure of electric vehicles as systems, and therefore the innovation process it-self. The engineers were acutely aware that they were primarily in the business of making things that "kind of work most of the time" and that above all had to

---

48 Russell, Andrew L./Vinsel, Lee: "Hail the Maintainers", in: Aeon, https://aeon.co/ essays/innovation-is-overvalued-maintenance-often-matters-more (accessed 06.09.2019); McCray, Patrick: "Mo's not all lightbulbs", in: Aeon, https://aeon.co/essays/most-of-the-time-innovators-don-t-move-fast-and-break-things (accessed 10.09.2019).

49 Vincenti, What Engineers Know, p. 200.

50 Andrew L. Russell and Lee Vinsel explicitly deplore that the history of technology has so far "focused predominantly on the earliest stages of technological life cycles", implying that maintenance is not (yet) part and parcel of technology at this stage. Consequently, they titled their paper "*After* Innovation, Turn to Maintenance", sug-gesting that one comes first and the other follows. Russell, Andrew L./Vinsel, Lee: "After Innovation, Turn to Maintenance", in: Technology and Culture 59 (2018), p. 1–25, here p. 4.

be just good enough to conform to real-world requirements, probably even more so than to be ground-breaking and exciting.[51]

To sum up, the case of 1970s electric vehicles suggests that understanding artefacts in all their complexity still challenges the available theoretical frameworks. Social and technological factors, invention and use, innovation and maintenance all matter simultaneously and relate to each other in ways that stark theoretical distinctions cannot always adequately capture. Historians of technology therefore have to tap into a number of different approaches simultaneously to grasp the complexities involved in working with artefacts, new or old. In this chapter, I have tried to bridge several such approaches, by considering questions of use and maintenance while describing an innovation process.

---

51 I am borrowing the slogan from "The Maintainers" group of researchers: http://the maintainers.org/ (accessed 04.04.2019).

REUSE AND CONSERVATION

A Bargain or a "Mousetrap"?

A Reused Penicillin Plant and the Yugoslavians' Quest for a Healthier Life in the Early Post-War Era

Sławomir Łotysz

In late 1947, an old penicillin plant that the Merck Corporation had successfully operated in Montréal for several years was dismantled and shipped to Yugoslavia to be re-erected in the empty building of an old textile plant in Zemun, a neighbourhood of today's Belgrade.[1] It was part of the so-called penicillin plant programme, an ambitious rehabilitation scheme that the United Nations Relief and Rehabilitation Administration (UNRRA)[2] had launched in January 1946. The programme included four other countries – Czechoslovakia, Poland, Belarus and Ukraine –, but unlike Yugoslavia they all received complete sets of brand-new factory equipment.

What might look like a striking example of inequality in the distribution of aid resulted from a sovereign decision by the Belgrade government. Initially, Yugoslavia was also offered new machinery, but it withdrew from the programme and entered into negotiations with Merck over its old plant. UNRRA, undeterred, upheld its commitment and paid for it anyway. It was only when the

---

1   This research is part of an ongoing study of UNRRA's penicillin plant programme and was funded by the National Science Centre, Poland, under grant number 2014/13/B/HS3/04951.

2   UNRRA was established in 1943 to bring assistance to victims of war and prepare for the reconstruction of war-torn countries when the hostilities were over. Offering penicillin plants instead of just sending periodic supplies of ready-to-use medicine reflects its main slogan "Helping people to help themselves". See United Nations Information Office: Helping the People to Help Themselves; The Story of the United Nations Relief and Rehabilitation Administration, New York: UN Information Office 1943, p. 13.

shipments of machinery were received that the equipment was found to be heavily worn out and incomplete. As a result, the Yugoslavians faced enormous difficulties putting the plant back together and bringing it into operation, all the while knowing that the final product would not meet contemporary standards of effectiveness in medical therapy, technological excellence and economic profitability.

In this chapter I will take a closer look at the transfer of the Canadian Merck penicillin plant that, despite its worn-out state and technical obsolescence, was finally put to work in Yugoslavia and produced penicillin for many years. By looking at the twists and turns in the story I will try to carve out the arguments and motives of the Yugoslavian actors who chose to acquire a second-hand plant instead of opting for new equipment. This historical case will also highlight the long "lifespan" and persistence of technical installations in process industries like drug manufacturing. Of course, the re-use of machines and even entire factories is common practice in industry worldwide. In fact, in certain industries, trading used capital goods is an important part of the business. As the renowned British economist John Maynard Keynes noted: "where the instrument is not irrevocably fixed to the ground, there generally is a second-hand market".[3] When a new product or technological process is introduced, old machines can be sold to another manufacturer, for whom they may still have potential value as a means of production.[4] And unlike modern equipment, they can be operated by less trained workers, and the commodities thus made – although not of top-notch quality or sophistication – can still be sold on less demanding markets. For buyers, taking such an option into account assumes a compromise between financial efficiency and technical performance, which can easily be calculated in an investment plan.

Ideally, it is a win-win situation: the seller can conveniently offset the costs of modernisation endeavours, while the buyer can make an investment at a fraction

---

3   Cited after Perelman, Michael: Keynes, Investment Theory and the Economic Slowdown: The Role of Replacement Investment and q-ratios, New York: Palgrave Macmillan 1989, p. 120.

4   Another reason for disposing of old factory equipment is the introduction of tighter environmental laws. Selling on to countries with less strict regulations can be tempting for many companies. For example, both economic obsolescence and stricter environmental regulations caused a large flow of used machines from Taiwan to continental China in the mid-1980s. See La Croix, Sumner/Xu, Yibo: "Political Uncertainty and Taiwan's Investment in Xiamen's Special Economic Zone", in: La Croix, Sumner/Plummer, Michael/Lee, Keun (eds.): Emerging Patterns of East Asian Investment in China: from Korea, Taiwan, and Hong Kong, New York: M. E. Sharpe 1995, p. 123–142, here p. 134.

of the cost.[5] In practice, however, it may also be a case of "lemons", according to George Akerlof's theory of quality and uncertainty in re-use practices.[6] In Akerlof's model of "cherries and lemons", capital or luxury goods offered for sale are more likely to be in bad technical condition than those which are not openly available but are sold among closer acquaintances or partners instead.[7] In this chapter I will consider whether the penicillin plant the Yugoslavians got from Merck was indeed a "lemon", or – to cite one of the World Health Organization officials – "a mouse trap",[8] or a "cherry" that helped to transfer penicillin technology from Canada to Yugoslavia.

The chapter is based on primary sources, mainly comprising archival records from Yugoslavian, foreign and international institutions involved in post-war reconstruction in Europe. The story of the implementation of the penicillin plant programme in early post-war Yugoslavia is absent from the international historiography concerning both UNRRA's activity in the Balkans and penicillin production.[9]

AN OFFER (NOT) TO BE REJECTED

The idea of establishing penicillin production in Eastern Europe originated in the summer of 1945, when the Czechoslovaks asked UNRRA to provide them with

---

5   Sometimes the economic value of used machines can be even higher than their book value. See, for example, Xue, Qi: Direct Foreign Investment, Technology Transfer and Linkage Effects: A Case Study of Taiwan, PhD dissertation, Case Western Reserve University 1979, p. 46.

6   Perelman, Keynes, p. 120; Akerlof, George: "The Market for 'Lemons': Asymmetrical Information and Market Behavior", in: Quarterly Journal of Economics 83, 3 (1970), p. 488–500.

7   Akerlof employed the example of second-hand cars to demonstrate the phenomenon whereby, because of the distrust of potential buyers, more expensive goods in better condition ("cherries") disappear from official circulation, while inferior ones – "lemons" – remain, causing even more distrust and leading to consequent price drops. Ibid.

8   This was Leslie Atkins, who supervised medical supplies at UNRRA and after its dissolution headed a similar division at WHO. His opinion is cited in a broader context later in this text (see footnote 51).

9   With the exception of the self-published account of a former employee of the first Yugoslavian penicillin plant from the 1960s which provides an eyewitness view of its development. See Bosnić, Petar: Istorija Jugoslovenskog penicilina 1945–1995, Belgrade: P. B. Bosnić 1995.

the means to make the drug at home instead of having to rely on deliveries from abroad.[10] UNRRA agreed to this and also extended the offer to other European nations, including Yugoslavia. Each package included the delivery of a complete set of factory-new technical equipment and machinery, the strains used to grow the *Penicillium* culture and the raw materials needed for six months of operation. The offer also included fellowships for two trainees from each country, a chemical engineer and a microbiologist, who would oversee the launch of the production process. All fellows were to be trained at Connaught Laboratories at the University of Toronto under the supervision of Norman L. Macpherson, the chief designer and manager of a plant operating at the lab. The blueprints that also came as part of the offer were drawn up based on this particular plant.[11] As soon as early August, Leo Rabinović, a medical adviser at the Yugoslavian Embassy in Washington, reported the news to Belgrade.[12]

For inexplicable reasons, however, Belgrade did not react for a couple of months,[13] and when it did, the Deputy Minister of Health, Grujica Žarković,

---

10 In fact, the Czechoslovak government in exile had investigated this opportunity even earlier, in 1944, but at that time their request to UNRRA was not accepted. For more details on the Czechoslovak initiative and how the programme was negotiated, see Łotysz, Sławomir: "International Health Organizations and the Dissemination of Penicillin Production Methods in the Early Cold War Era. The Case of the United Nations Relief and Rehabilitation Administration Activities in Europe, and the Work of the American Bureau for Medical Aid to China in China" [in Chinese: Guójì wèishēng zǔzhī yǔ lěngzhàn chūqí pánníxīlín shēngchǎn fāngshì de chuánbò - yǐ liánhéguó shànhòu jiùjì zǒng shǔ zài ōu huódòng hé měiguó yīyào yuán huá huì zài huá shìwù wéi lì], in: Journal of the Social History of Medicine and Health [Yīliáo shèhuì shǐ yánjiū], 2, 1 (2017), p. 3–31; id.: "Knowledge as Aid: Locals Experts, International Health Organizations and Building the First Czechoslovak Penicillin Factory, 1944–49", in: Reinisch, Jessica/Brydan, David (eds.): Europe's Internationalists: Rethinking the History of Internationalism, London: Bloomsbury 2021, p. 140–157.

11 The aims of the programme are outlined in various texts, including: Łotysz, Sławomir: "A 'Lasting Memorial' to the UNRRA? Implementation of the Penicillin Plant Programme in Poland, 1946–1949", in: ICON: Journal of the International Committee for the History of Technology 20, 2 (2014), p. 70–91.

12 Ministarstvo inostranih poslova: Letter to Ministar trgovina i snabdevanje, 6 Aug. 1945, 671/11, Arhiv Jugoslavije, Belgrade, Serbia (hereafter: AJ).

13 One may assume that this delay was caused by the political turmoil prevailing in Belgrade at that time. Since defeating the German occupants, the communists of Marshal Tito had exercised real power, but it was only in November 1945 that the parliamentary election legitimised the post-war state of affairs in Belgrade. Significantly, the ministry replied a day after the election.

responded that Yugoslavia "had never given its consent" to UNRRA buying a penicillin plant on its behalf, and more or less openly refused to take part in the programme.[14] In fact, the ministry was afraid of the costs involved in launching production of antibiotics. According to the deputy minister, such a factory would cost them at least one million US dollars (although he did not specify the source of these calculations). To that amount he added another 200 million dinars – or 4 million US dollars at the official exchange rate – for assembling the factory after its components were shipped to the country.

Žarković said that such a fortune could be used "more effectively for purchasing other medical equipment or even ready-to-use penicillin".[15] The ministry was also concerned that the plant's output was "significantly" higher than the country's actual needs. If such a factory was erected in Yugoslavia, Žarković argued, it would "have to export" surpluses of the produced drug, and the deputy minister had little faith in Yugoslavia's chances of competing on global markets. In this way, he also seemed to be suggesting that "such a factory" fell some way below the highest standards. Since further improvement in antibiotic production was widely predicted, especially with the advent of synthetic penicillin, then this gap would only increase even more.[16]

The arguments given by Žarković were as bizarre as they were unclear. Apart from the fear of what to do with the excess amount of antibiotic produced, the calculations quoted were significantly overstated, as if in deliberate opposition to the plant. The costs given in the minister's letter were three times higher than those specified in the UNRRA estimates. Besides, all expenditure incurred in the procurement and shipping of the equipment was to be covered by UNRRA – something the letter did not indicate that the Ministry of Health had understood.

On 30 January, just a few days after the programme was officially announced, the Yugoslav government officially notified the UNRRA mission in Belgrade that it did not need a penicillin factory as well as a hospital and medical supplies.[17] Michail Sergeichic, the head of the UNRRA mission in Belgrade, explained the position of the local government: "This has only been done to have money with which to buy food, which must be kept moving into Yugoslavia

---

14 Pomocnik Ministra narodnog zdravlja: Letter to Zavod za vanredne nabavke, 27 Dec. 1945, 671/11, AJ.

15 Ibid.

16 Ibid.

17 UNRRA Belgrade: Cable to UNRRA Washington, 3 Feb. 1946, S-1443-0000-0056, United Nations Archives, New York City, USA (hereafter: UNA).

without interruption."[18] On 11 February, UNRRA offered the Italian government the opportunity to participate in the programme, and it gladly accepted.[19]

Additional reasons for Yugoslavia's withdrawal from the programme can be partly explained by Vladimir Kušević, a director of the General Board for Medical Production (Glavna Uprava Medicinkske Proizvodnje – GUMPRO). He went to America in May 1946 to oversee the investment purchases needed for the reconstruction of the Yugoslavian pharmaceutical industry.[20] During his three-month stay, Kušević regularly visited pharmaceutical and chemical companies throughout the United States. In August he appeared at the medical department at the UNRRA headquarters in Washington.

Whilst there, he was asked why Yugoslavia had passed up an offer that everyone else had immediately accepted. He explained that the perception in Belgrade was that "penicillin plants were vastly complicated affairs, requiring over 400 people to operate", which – as the chargé d'affaires of the Canadian Embassy in Washington commented – was "apparently too much even for the vigorous Yugoslavia".[21] Kušević hinted that his government "had been deliberately misled by some dark commercial interests who, presumably, had hoped, at a later date, to go into private and profitable penicillin production".[22] What exactly these forces were, he did not disclose.

Eventually, after their doubts had been dispelled, the government in Belgrade requested reintroduction to the programme. The UNRRA staff, "undeterred by the rather fatuous explanations", immediately began preparations for the purchase and shipment of the factory equipment. Time was running out because, according to UNRRA's schedule, all orders had to be submitted by 1 October 1946

---

18 Sergeichic, M.: Cable to I. Sollins, 4 Feb. 1946, S-1412-0000-0068, UNA.
19 The plant was shipped to Italy and erected there, but it was launched with much delay only in 1952, mainly because the Italians wanted to couple it with a large biotechnology research centre. For information on how they did that, see, amongst others: Capocci, Mauro: "'A Chain is Gonna Come'. Building a Penicillin Production Plant in Post-war Italy", in: Dynamis 31, 2 (2011), p. 343–362; Cozzoli, Daniele/Capocci, Mauro: "Making Biomedicine in Twentieth Century Italy: Domenico Marotta (1886–1974) and the Italian Higher Institute of Health", in: The British Journal for the History of Science 44, 4 (2011), p. 549–574.
20 Monograph on the medical and sanitation program for Yugoslavia, undated, p. 5, S-1021-0014-21-35, UNA.
21 Canadian Embassy in Washington: Letter to Secretary of State for External Affairs, 7 Sep. 1946, RG 25-3798-8286-40, Library and Archives Canada, Ottawa, Canada (hereafter: LAC).
22 Ibid.

and all deliveries completed by the end of the following March. This strict time-
line was demanded by the United States, which was the main supplier of industrial commodities within UNRRA's various programmes. As could be seen from
the progress made by Czechoslovakia and Poland in the penicillin plant programme, some US manufacturers were unable to meet the orders sooner than
October 1947. Nevertheless, the plant was added back to fiscal estimates for
1946 supplies to Yugoslavia as a $400,000 "additional request". This plant was
basically the same design as the one Belgrade had passed over before.[23]

Initially, Yugoslavia contemplated negotiating with the United States for an
extension of the delivery deadline by several months, preferably to the end of
1947. There was even a precedent for this: the delivery period had been extended
for fuels, food, spare parts and materials of particular importance in the post-war
reconstruction of the country. The penicillin factory could also be considered
particularly high priority. Eventually, though, most likely assuming that the Department of State would not agree to such an extension, in the second half of
November 1946 the Yugoslavians again cancelled their participation in the programme.[24]

YUGOSLAVIA GOES OUT ON ITS OWN

The Yugoslavians had by no means given up on their efforts to obtain a penicillin plant. Instead of counting on the continuity of supplies after the dissolution of
UNRRA, however, they sought to purchase a used factory on the free market. At
the end of January 1947 the Federal Commission for National Health (*Komitet
za zastitu narodnog zdravlja* – KZNZ) sent Josip Milunić, a doctor of medical
sciences from Zagreb and a loyal communist, to the United States to find the
best offer.[25]

It is not known exactly when and where Milunić began negotiations, but by
early March he had already reported from Washington on quite advanced talks
with the Merck Corporation about its old penicillin plant still operating in Mont-

---

23 Garfield, S.: Cable to K. Sinclair-Loutit, 27 Sep. 1946, S-1414-0000-0553, UNA.
24 UNRRA Washington: Cable to UNRRA Belgrade, 20 Nov. 1946, S-1443-0000-0057,
    UNA.
25 KZNZ: Letter to Predsedništvo Vlade FNRJ, 25 Mar. 1947, 31/60/87, AJ. The infor-
    mation on securing financial resources for this purpose was taken by KZNZ no later
    than 20 February, see: KZNZ: Letter to Ministarstvo Spoljne Trgovine, 20 Feb. 1947,
    31/60/97, AJ.

réal.[26] If Yugoslavia accepted the offer, for half a million US dollars the factory would be dismantled, shipped to Europe and then reassembled. Milunić mailed Belgrade all the technical data he had received from Merck, along with detailed dimensions of the factory building, which had to be erected or adapted for the purpose on site. In order to get acquainted with the layout and operation of the plant while still in use, he scheduled a visit to Montréal in early March. Milunić planned to travel with Ivan Radenović, a chemical engineer working as a technical adviser to the Yugoslavian Embassy in Washington.

Milunić reported that Merck's offer included the training of two Yugoslavian specialists, a bacteriologist and a chemical engineer, who would be tasked with setting up the factory. In addition, the company promised to send in three of its own specialists to help launch production. He had no doubt that it was an attractive offer, and even the evidently poor condition of some of the devices, as well as clear signs of alterations to the piping and other instruments, did not bother him. After all, it had been "one of the first penicillin plants in America", as he explained.[27]

The first things to be replaced after bringing this plant over to Yugoslavia were the fermentation tanks. The existing Montréal tanks were 700 gallons each, while most modern factories were equipped with rows of tanks that were three times more capacious. Switching to larger containers had many advantages, such as being able to adjust the fermentation section to make streptomycin. According to Milunić, three large tanks would have to be put up once the factory was reassembled in Yugoslavia.

Other sections of the production line, particularly the centrifuges and dosing apparatus, were also in poor condition and desperately needed replacement upon reassembly. In addition, some of the technical solutions employed at Merck's plant were regarded by Milunić as being "atypical". For example, the drying

26 Milunić, J.: Cable to KZNZ, 1 Mar. 1947, 31/60/87, AJ.
27 Id.: Report to KZNZ "Izveštaj o fabrici penicilina", 11 Mar. 1947, p. 2, 31/60/87, AJ. Interestingly, a CIA report from November 1953 refers to the transferred plant as "the first in the world to produce penicillin" (see Yugoslav research in pharmaceuticals / Need for technical know-how, 12 Nov. 1953, CIA-RDP82-00047R000300510007-4). In fact, the Merck factory in Montréal was the third – after Connaught Labs and Ayerst, McKenna & Harrison Co. Ltd – to make penicillin using the surface culture method in Canada, in summer 1943. Feasby, W. R. (ed.): Official History of the Canadian Medical Services 1939–1945, vol. 2, Ottawa: National Defence 1953, p. 391. By July 1944, the installation had been converted to use the deep fermentation method, the first of its kind in the British Empire. Warrington, Charles J./Nicholls, Robert van: A History of Chemistry in Canada, New York: Pitman 1949, p. 291.

process utilised an oil vacuum pump, while the best results could be obtained with steam pumps.

The half a million dollars that Merck wanted for the plant was more or less the value of two months' production, at least according to the company itself. Given the technical condition of the factory, Milunić considered the price too steep. In his opinion, the actual value of all the equipment was about $100,000. He also thought that the plant was priced so steeply because the company wanted to cover the profits it would lose after it was dismantled.[28] On the other hand, accepting this offer would have made the Yugoslavians self-sufficient in terms of penicillin supplies in the shortest time possible. At that time it was already evident that the deliveries of UNRRA equipment to Czechoslovak and Polish plants were experiencing difficulties, and they would not open as swiftly as initially hoped. Milunić therefore encouraged his superiors to seriously consider the matter.

According to Merck, under their management the plant produced 50 billion Oxford units of penicillin every month, one unit equaled 0.6 micrograms of crystalline compound. They made amorphous antibiotic, which was then purified to pure crystals. It was the latter fact that most appealed to the Yugoslavians in the offer. The UNRRA plant employing technology developed at Connaught Labs could only make penicillin in amorphous form, which was inferior in terms of curative power. The capacity and economy of production were similar in both cases. At the Merck plant, the production cost of 100,000 units was 18 cents, which did not differ much from the average for plants of that size. The cost of raw materials constituted some 10% of this amount, and altogether this suggested a very profitable enterprise.

Anticipating possible questions about why, then, the company was willing to get rid of such a seemingly lucrative gem, Milunić informed Belgrade that Merck planned to build a much larger penicillin and streptomycin factory in the area. This new undertaking was part of the company's strategy to keep up with changing trends in the antibiotics industry in North America. As a result of the sharp increase in demand for penicillin in the final part of the war, many new factories were built in the United States and Canada, leading to significant overproduction of the medicine and thus fierce competition on the market.

The subsequent price drop necessitated further concentration of production and cost-effectiveness measures to be taken, and in the long run smaller factories were unable to compete. To keep up with their competitors, Merck had to build a factory with a significantly greater production capacity than the one in Montréal.

---

28 Warrington/Nicholls, A History.

The old machinery still had some value, though, and its production capacity was several times in excess of the demand of an average country, so why not sell it?

Along with Merck's proposal, Milunić presented another from the University of Toronto, which was a repeat of the Connaught Labs model that Yugoslavia had rejected twice already. He was ostensibly leaning toward the first option. Milunić argued that they should hurry up if Merck's offer were chosen, as other countries were supposedly now queuing up to take the plant home (although he didn't explain who these eager competitors were).[29] It is not known now whether there were actually any rivals for the plant, or whether was this just a bluff to urge the Yugoslavians to make their minds up faster. But either way, they finally agreed and Merck got the deal.

Yugoslavia allocated $700,000 for the purchase.[30] Dmitar Nestorov, chairman of the KZNZ, and its secretary Voja Djukanović kept pushing the government to hurry up and finalise the deal. They argued that the versatility of penicillin meant that it replace many imported drugs. But their crowning argument was ideological in nature: they emphasised the importance of the plant's purchase to the country's five-year economic plan, saying that "the use of penicillin would save and quickly bring back to work tens of thousands of workers, and the national economy would not suffer from a lack of manpower".[31]

When, sometime in late April, information about the planned deal reached UNRRA, it offered to cover the Yugoslavians' expenses. It is not known whether Belgrade asked for this, or if the Administration saw a chance to fulfil its commitment to provide the Balkan nations with their own penicillin, but it did secure half a million dollars for the purpose. On 28 April, the head of the Medical and Sanitation Supplies Division at the Washington headquarters, Irving V. Sollins, went to Canada to settle the details. Formally, the negotiations were carried out by the Canadian Commercial Corporation, which handled local purchases for UNRRA.

Somehow, though, while visiting Washington in mid-June, a certain Mr Low from the Canadian Commercial Corporation informed the Canadian Ambassador that "by refusing to be rushed into acceptance of an absurdly high price, they had been able to obtain a reasonable offer of $225,000 from the Merck Company".[32] Because of this misunderstanding, the deal was put on hold for another three

---

29 Ibid., p. 3.

30 KZNZ: Letter to Predsedništvo Vlade FNRJ, 25 Mar. 1947, 31/60/87, AJ.

31 Ibid.

32 Canadian Embassy in Washington: Letter to Secretary of State for External Affairs, 3 Jul. 1947, RG25-3798-8286-40, LAC.

weeks until the formal acceptance of the government in Ottawa was finally received on 3 July. The contract was then signed two weeks later.[33]

TAKING IT DOWN AND PUTTING IT BACK UP AGAIN

The contract negotiated by the Yugoslavians was not only overpriced. Although its exact terms are not now known, from the way in which it was implemented it is apparent that it put the buyers in a very disadvantageous position. When in August the KZNZ sent in another of its technical experts, Novaković, to oversee the disassembly process in Montréal, the factory management would not let him in, "so he could not see the devices working".[34] He was admitted only on 5 September, when all of the control devices had already been taken down and actual disassembly of the production lines had been started by a contracted engineering firm, the Donald & Ross Company. All pieces of equipment were carefully marked and identified on a diagram that was to be included with the shipments. The process of actually sending the thus boxed-up factory to Yugoslavia was UNRRA's responsibility.[35]

Two weeks later, yet another Yugoslavian representative arrived in Montréal. This was Dr Miho Piantanida, a biochemist who headed the hormonal preparations department at the Pliva plant in Zagreb.[36] He was an eminent figure in the Yugoslavian medical world, having been the first to extract domestic insulin, which was then introduced into medical practice in 1940.[37] If he had come to Montréal to familiarise himself with the factory, he was definitely too late – there was not much of the machinery left in the building. And he was not accompanied by any microbiologists, so only half the team needed to get the plant up and running again was ultimately present.

Piantanida did not wait to oversee the shipment of the factory parts and flew back to Yugoslavia. Once there, he discovered that while he had been in Canada, the authorities had changed their minds as to where to set up the factory. The ini-

---

33 Zavod za Vanredne Nabavke, Ministarstvo Spoljne Trgovine: Letter to KZNZ, 3 May 1947, 31/60/87, AJ.

34 Anon.: "Postrojenje za proizvodnju penicilina (Pro memoria)", undated, 83/07, AJ.

35 Piantanida, M.: Letter to GUMPRO, 28 Nov. 1947, 31/60/87, AJ.

36 Anon.: "Postrojenje".

37 Labai, Boris, et al.: "Događaji koji su mijenjali hrvatsku medicinu. Razvoj i postignuća u struci i znanosti", in: List Medicinskog Fakulteta, 36, 2 (2017), p. 9–28, here p. 13.

tial intention was to build a new building not far from Belgrade, but after realis-
ing that the necessary preparatory work at that location would have taken at least
two years, KZNZ decided to set up a temporary plant instead. Still, before de-
parting for Canada, Piantanida suggested that the company's facilities in Pliva
would be the best choice, as Pliva had both available factory space and a skilled
workforce. When he returned to the country, however, the authorities showed
him an old factory that had once housed a textile company in Zemun, on the outskirts of the capital city. So when, at the end of December 1947, the
SS *Marchport* unloaded its precious cargo in Trieste, it was immediately sent by
train to Zemun, where it was deposited in a warehouse at the designated site.[38]

The reassembly works at Zemun began in February 1948. Piantanida was
confident of his ability to set up the plant with his team only and insisted on not
hiring any foreign experts.[39] In the strongly idealised narrative of the national
press, the lack of knowledge and experience was replaced with revolutionary
zeal. It was reported that on the first day of work, "the engineers, technicians and
workers [had already] collectively made a commitment" to complete the construction by Marshal Tito's 57th birthday on 7 May 1949.[40] But from the very
beginning, the reassembly did not go smoothly. In the spring of 1949, Piantanida
reported to the KZNZ that although immediately after his arrival from Canada he
had given his opinion on the degree of wear of the plant and its essential obsolescence, during the assembly these concerns had not only proved accurate; the
situation was actually much worse than was possible to determine at first glance
during the hurried disassembly process.[41] He asserted that setting up the plant at
Zemun would require a significant financial outlay. A large part of the equipment had to be replaced, but "as there was practically no single piece of machinery that was not worn out, an overhaul would mean actually replacing all the
equipment", which he did not consider justified. For these reasons, he recommended building a new factory instead of "patching up the old one, which in no

---

38 Anon.: "V naši državi bomo zgradili tovarno za penicillin", in: Ljudska Pravica
(Ljubljana), 30 Dec. 1947, p. 3.
39 Anon.: "V Jugoslaviji so začeli izdelovati čudovito zdravilo – penicilin", in: Enako-
pravnost (Cleveland, USA), 25 Oct. 1949, p. 2. Piantanida's team included, amongst
others, microbiologist Gavra Tamburašev, pharmacists Živka Pešič and Slavica Mir-
kovič, chemical engineers Z. Peric, A. Jovič, and A. Sekulič, and technologist B. Pav-
kovič. An architect named Božidar Petrović adapted the building at the old factory in
Zemun for its new purposes.
40 Anon.: "Začela je obratovati tovarna penicilina", in: Ljudska Pravica, 26 May 1949, p. 2.
41 Unsigned report to KZNZ, undated, 31/08/22, AJ.

case could meet contemporary requirements in terms of both efficiency and profitability, as well as the quality of the final product".[42]

Piantanida was convinced that the monthly production capacity of 50 billion units, as declared by Merck, was based on the purely theoretical assumption that the concentration of active substance in the fermentation broth fluctuated around 1,000 units of active substance in one cubic centimetre of broth. He believed that such a yield was simply impossible to reach due to the faulty design of the plant. Upon experimenting, he found that the aeration system mounted in the fermentation tank was insufficient for the volume of medium processed in one batch. Piantanida was of the opinion that apart from the heavily worn-out condition of the factory, this was one of the main reasons why the company wanted to get rid of it. He put it frankly: "If it really could have competed with other large companies in America, it's hard to believe Merck would have put an end to it".[43]

His view was confirmed by foreign experts, particularly Ernst Chain, whom he met at a conference in Geneva on 17 February. The meeting was jointly called by the World Health Organization and the Economic Commission for Europe, to discuss ways of assisting the beneficiaries of the former UNRRA (after it was closed down, WHO took over its unfinished health programmes) in completing their penicillin plants. The Czechs and Poles, who had already encountered serious problems with their own assembly processes, warned the international organisations that further delays in launching the plants might endanger implementation of their ambitious anti-venereal campaigns, which had been designed on the assumption of an abundance of domestic penicillin.

Piantanida went to Geneva with a microbiologist, Gavra Tamburašev, with whom he worked on reassembling the plant.[44] They examined the plans of the factory as well as photographs of the technical equipment that had already been installed at the Zemun plant. As Piantanida reported, the experts asserted that it was "completely useless to re-install this outdated plant".[45] This opinion was included in the conclusions of the conference. WHO promised to assist Czechoslovakia, Poland and Yugoslavia in bringing the production methods employed at all three factories up to date. This meant, amongst other things, providing them with Podbielniak extractors to modernise the plants' extraction departments.

---

42 Ibid.
43 Ibid.
44 Tamburašev was made a director of the plant and is now widely known as a "father of Yugoslavian penicillin". Bosnić, Istorija, p. 61.
45 Ibid.

While analysing Piantanida's report of the meeting, the deputy director of GUMPRO, V. Pavlov, acknowledged the fact that the Yugoslavian factory was "actually the oldest" of all three plants discussed in Geneva and that production there would be "most likely unprofitable", at least by American standards. "This fact was known to representatives of UNRRA when they offered this factory, and also to us, when we received it."[46] But in summarising his grievances Pavlov did not mention that it had been Milunić who had negotiated the deal.

Ultimately, though, Pavlov recommended that the Ministry of Light Industry should continue construction work at Zemun, regardless of whether or not WHO kept its promises. Any shortcomings in the equipment and on the economic side of the enterprise could, according to Pavlov, be assessed only after the factory had entered operation. This was his indirect answer to the report's conclusion, which was to reconsider the practicality of further assembly work at Zemun.[47]

THE "MERCK MOUSETRAP"

In early May 1949, just a few days before Tito's birthday, assembly of the plant was completed and start-up trials began.[48] The first antibiotic sample was obtained on 31 August.[49] Although the quantity was so small as to be enough only for research purposes, the press promised the "imminent" start of full-scale industrial production. The trial run lasted until the end of September, when the workers felt that they were expected to make another grand commitment for the advancement of the party. This time they promised to run the factory in top gear.[50] This news raised hopes in Yugoslavian society that penicillin would become more widely available on the market in the coming months.

When information about the trial run at the plant came out in the West, Leslie Atkins, then head of the WHO Purchasing Division, said:

"One of their big shots gave a long lecture reciting how pharmaceutical production had increased. Says he, basing his figures on production for 1945, 'The country's production has

---

46 Pavlov, V.: Letter to Ministarstvo Lake Industrije FNRJ, 8 Mar. 1949, 10/55/57, AJ.
47 Ibid.
48 Anon.: "Začela", p. 2.
49 Anon.: "Naš delovni kolektiv je obvladal tehnološki proces proizvodnje penicilina", in: Ljudska Pravica, 7 Sep. 1949, p. 3.
50 Anon.: "Proizvodni plan za mesec avgust je večina tovarn in direkcij zvezne lahke industrije presegla", in: Ljudska Pravica, 7 Sep. 1949, p. 2.

increased progressively 3,000%.' Well, with 1945 and all its troubles as a base year, my comment is 'Twice nothing is still nothing.' It will be very interesting to see what they can do with the Merck mousetrap."[51]

The opinion expressed by Atkins best illustrates the actual value of Merck's offer, but it did not change the fact that the Yugoslavians had shown enormous determination in getting into this "mousetrap" of their own accord. The fundamental mistake on their side was to accept an offer that did not include the transfer of know-how and did not even allow the Yugoslavians to assist with the disassembly of key elements of the machinery. And despite Milunić's assertions, the Yugoslavian specialists who went to Canada had not been trained.[52] The fact that the plant's devices were obsolete had been known from the beginning, but the extent to which they were worn out still surprised the buyers.

The Yugoslavian authorities blamed UNRRA and unspecified "dark commercial forces" for what had happened, but not themselves. They had even neglected the warning signs from their own negotiators sent to the United States and Canada as purchasing agents, which is what had caused all the problems in the first place. They had agreed to pay half a million dollars for the plant, despite having been warned by Milunić that the price was too steep considering the plant's condition. Only when UNRRA stepped in did Merck reduce its expectations to $225,000. Still later, when the assembly process at Zemun came unstuck, the Yugoslavians realised that it was way too much for a production line built for $40,000 during the war, and amortised several times since then. Considering its current condition and usability, they estimated that its actual cost was closer to $20,000.[53]

What was worse, when the shipments arrived it turned out that the equipment was incomplete. The entire freeze-drying section was missing, as well as a steam boiler and a machine for hermetically sealing ampoules of penicillin. As for the latter device, the signed contract had an annotation on it stating that, according to Milunić, it was not needed.[54] But it definitely was. In early January 1950, at a

51 Atkins, L.: Letter to N. Macpherson, 30 Jun. 1949, 83-016-04, Sanofi-Pasteur Archives, Toronto, Canada (hereafter: SPAT).

52 The actual text of the agreement was not accessed, and therefore it cannot be determined with certainty whether Merck did not fulfil the contract or whether the provisions in the contract differed from the initial promises made to Milunić.

53 This estimate comes from a memo, the physical document of which is incomplete. The remaining first page of the document is neither signed nor dated. See Anon.: "Postrojenje".

54 Ibid.

special conference held in Belgrade to look for a way out of the deadlock, the new president of KZNZ, Pavle Gregorić, forgave the staff members who had been generally criticised for failing to launch penicillin production as they had promised in their commitment the previous year: "They could not come up with something that was impossible. They were not provided with even basic conditions to work, and they could not perform a miracle".[55] Gregorić even complemented them, saying that they had given "their best efforts and achieved a truly unexpected success".[56]

By "success" he was referring to their mastery of the fermentation process. At Zemun they had indeed managed to culture *Penicillium* fungi and extract salts of penicillin from the fermentation broth. However, at that point they had had to stop the procedure because of the missing freeze-drying apparatus. Growing cultures in large tanks is always a complicated process, largely dependent on meticulous compliance with procedures and the experience of staff. But the measure of their success here was the fact that the Yugoslavians had to work under very tough conditions. Out of 38 fermentation batches they made during the trial run, only half were actually successful. In eight cases the entire batch degenerated owing to a failure of the electric installation, and in several others because of an inadequate water supply.[57] Piantanida had reasons for bitter satisfaction – even before the equipment reached Yugoslavia, he had warned GUMPRO that neither the water nor the electric power supply at Zemun would be sufficient, and again recommended setting up the plant in Pliva. However, as Gregorić had reminded him, at that time all of those warnings were rejected by "the comrades" in Belgrade.[58]

Piantanida was also right in his prediction that all attempts to modernise the installation at Zemun to supply crystalline penicillin would be useless. In fact, he had been saying this since the very beginning of the assembly process. It was already 1950, and the entire world was using only this form of antibiotic, which was more effective and much easier to use and store (unlike amorphous penicillin, which had to be kept at a low temperature and lasted only a few weeks before losing its potency).

---

55 Anon.: "Zapisnik konferencije održane u ministarstvu lake industrije FNRJ dana 7 I 1950 godine pro predmetu proizvodnje penicilina u tvornici u Zemunu", p. 3, 10/55/ 57, AJ.

56 Ibid.

57 Ibid., p. 4.

58 Ibid., p. 3.

Pavlov explained that when the decision to buy a second-hand plant had been made, the authorities in Belgrade believed that since amorphous penicillin had been in use since its introduction in medical practice in 1943, it would last for many years to come. It was only after the deal with Merck was signed that the management of GUMPRO saw the logic and practicality of shifting toward crystalline penicillin.

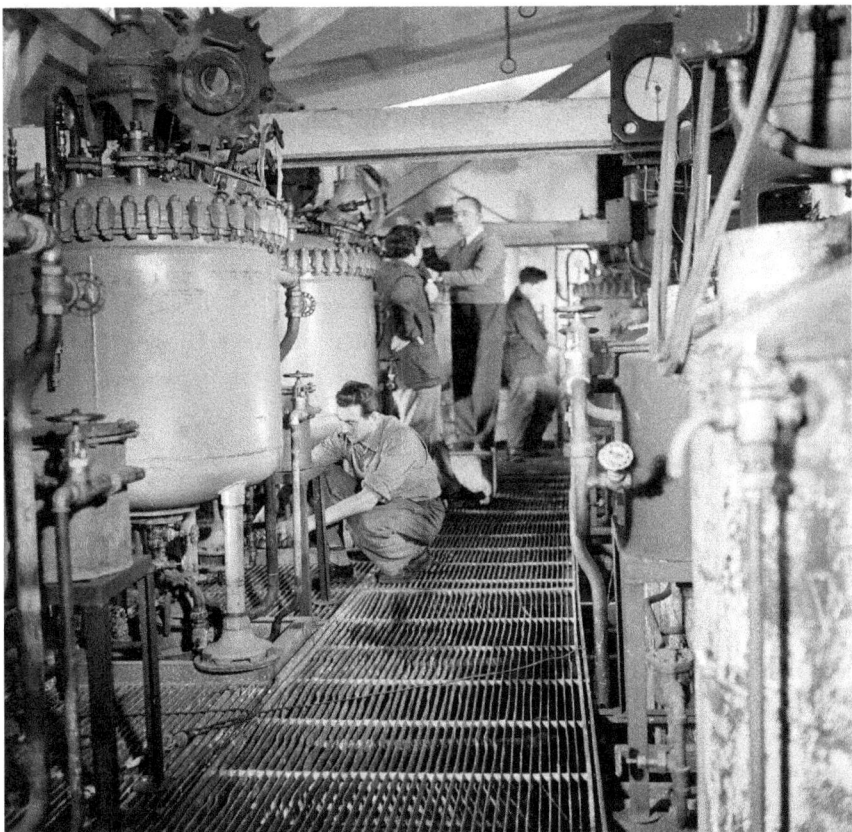

Figure 1. Upgrading the penicillin plant in Zemun in late 1953. The small tanks, in which a mixture is prepared for final fermentation in large tanks, were also part of the original Merck factory (Courtesy of the United Nations, UN Photo/GG, no. 156110.).

In many ways, at the beginning of 1950 the Yugoslavians no longer had any doubts that the plant was old and primitive, but they still tried to rationalise their erroneous decision to purchase it. It was argued that as a "pioneer among all penicillin factories in Canada", this plant had helped give experience to and train local specialists, and after being transferred to Zemun, it would play a similar role in Yugoslavia.[59] The January meeting concluded with a strong determination to complete the plant in Zemun, lest all efforts made so far be rendered futile. Realisation of the project still lagged behind, however, and industrial production was launched only in late 1950.

Aware of the shortcomings of its plant, the Yugoslavian government requested advice from WHO on how to improve it. As early as December 1950, WHO sent Macpherson, in his capacity as a penicillin expert, to survey the Zemun factory. Contrary to what Piantanida had claimed, Macpherson reported that the plant could easily be upgraded to produce antibiotics in crystalline form, but the cost of the modernisation would be around $90,000. Moreover, after the expansion and modernisation that he proposed, the plant would be able to make 30 billion units of crystalline penicillin per month – less than its nominal capacity of 50 billion units, since antibiotic crystallisation causes a loss in potency compared with its amorphous form. WHO arranged with UNICEF to pay for the new equipment and modernisation works at Zemun from residual UNRRA funds. Eventually, after a lengthy reconstruction period, on 8 January 1954 it began making crystalline penicillin at an even greater monthly capacity of 100 billion units (see fig. 1).[60] Despite these modernisation works, as well as two subsequent upgrades in 1958 and in 1966, the core of the factory was left more or less intact until 1973, when a completely new plant was erected in different location.[61]

CONCLUSION

The transfer of the penicillin plant from Canada to Yugoslavia is a startling example of technological persistence in its literal meaning as the persistence of form of a material object. It can be explained by the steadiness of the industrial process embedded within it. The example of Merck's old penicillin plant is even more significant in that the antibiotic industry made major advances in the late 1940s and early 1950s – production levels boomed, product quality increased and prices nosedived. This was partly because in principle, the process of bio-

---

59 Ibid., p. 6.
60 Anon.: "Recommendation of the Executive Director for an Apportionment. Yugoslavia. Penicillin Production Plant", 2 Aug. 1955, p. 1–2, E/ICEF/L. 785, UNICEF, https://digitallibrary.un.org/ (accessed 19.12.2014).
61 Bosnić, Istorija, p. 248.

synthesising penicillin had remained unchanged since Florey, Heatley and Chain defined it in the early 1940s.

Yugoslavia's decision to base its antibiotic industry on used machinery was not an attempt to compromise between financial efficiency and technical performance, as is usually the case. It was the result of a combination of factors, primarily lack of experience and ideologically driven distrust of international relief organisations. In other words, Yugoslavia was manoeuvred into this deal by the very "commercial forces" it was so afraid of. I have argued that the main reason the Yugoslavians were tempted into accepting Merck's offer was the appeal of having their own crystalline penicillin in a shorter time than if they accepted a plant from UNRRA. But their calculations failed for a number of reasons, such as substantial existing wear and tear on the machinery and inadequate training of technical personnel. On the other hand, the Poles and Czechs also struggled with their plants. Deliveries of equipment were irregular and also incomplete, which substantially delayed the launch of production. Poland made its first amorphous penicillin in July 1949, and Czechoslovakia three months later. The antibiotic in crystalline form came even later, during in 1952 in both countries.

By purchasing a second-hand penicillin factory the Yugoslavians gained access to a technology previously inaccessible to them, which in the end contributed to the establishment of an entirely new branch of the country's pharmaceutical industry, as well as to the emerging discipline of biotechnology. Admittedly, this could also have been achieved by building a new factory. One could argue, however, that the more the local specialists had to tinker to make the old equipment work, the more effective their training was.

Assuming that the Akerlof theorem applies also to investment goods, was the Merck plant actually a cherry rather than a lemon? Was it a bargain or a "mousetrap"? The market of second-hand penicillin plants was rather limited in the late 1940s, and Merck's negotiators made the Yugoslavians believe that by overpaying they would get the best offer before others would. It certainly appeared to be a cherry, until it was found to be dramatically worn out and incomplete. But it did eventually work out, and the plant operated for nearly three decades, effectively supporting Yugoslavians' quest for a healthier life. Today, the old Merck plant still persists in the public memory in Serbia – a commemorative plaque indicates where the Zemun factory once stood, while the Museum for the History of Pharmacy in Belgrade has a small exhibition featuring the story of penicillin production in Yugoslavia. In the exhibition, a statutory plate from a fermentation tank, presumably the last piece of the original equipment sent from Canada in 1948, offers tangible proof of the material persistence of technological objects.

"Proof of Life":

Restoration and Old-Car Patina

David N. Lucsko

In November 1986, Victor Ofner of Escondido, California, took delivery of a worn-out German-market VW. Delivered in Mannheim in the autumn of 1953, the grey-blue Beetle had plied the streets of the Federal Republic for twenty-six years before being placed in storage. There it sat until a friend of Ofner's found it and helped arrange its purchase and shipment to the US. After taking delivery of the inoperable but largely original car, Ofner spent the next eight-and-a-half years of his spare time on a thorough restoration. He rebuilt the engine and transmission, completed the bodywork, applied the paint, powder-coated the chassis, cad-plated every nut and bolt, and spent countless hours hunting for period-correct accessories. The outcome was flawless. The car took "best of show" when it debuted at a vintage meet in 1995, and in 1996 it was featured in *Hot VWs*. Ofner "outdid himself with this restoration", the magazine gushed, for his car was "now better than when it rolled off the factory assembly line".[1]

Ofner's results were exceptional, but his goals and methods certainly weren't. For many decades, from its modest origins in the interwar years to its heyday as a multifaceted mass phenomenon in the 1970s–1990s, the hobby of old-car restoration was largely predicated upon a widely-shared aim: like-new form. This goal was also shared by others in the broader old-car world: museums, high-end private collections, auction houses, insurance firms and the like. With the important exception of museum specialists, however, this essay focuses specifically on the like-new mantra as it played out among old-car hobbyists, the largely do-it-yourself crowd of middle-income gearheads who restore, drive and

---

1   Smith, Robert K.: "Above-Standard Standard", in: Hot VWs (HVW), Aug. 1996, p. 40–43, here p. 43.

show old cars in their spare time.[2] This is certainly not to suggest that these enthusiasts have always agreed on what exactly "like-new" means – or, by extension, what a "restoration" actually entails. Does it require a strict adherence to factory specifications, or are some modifications acceptable? Should drivability or durability be factored in? Should a car be rebuilt to the highest possible standard, like Ofner's Beetle, or should one aim instead to replicate the typically more modest standards of factory fit and finish? Finally, which cars ought to be restored – only high-end makes? Only rare examples? Only those of a certain age?

On these and other points, disagreement has long been rife. Yet across time, space and its many niches and sub-niches, the hobby of old-car restoration entailed at the very least an *effort* to erase all signs of wear and tear. Many never quite finished the job, of course, and "work-in-progress" cars have always been common at shows, accompanied by a litany of explanations with which every old-car nut could sympathise: "I hope to get it painted this autumn", "I'm still looking for the correct tail lights", "I might have finally found the right material for the seats". In short, not every restored car was actually perfect, but perfection was the aim.

By the early 2000s, however, a curious twist was in the making. Evident at old-car shows and in magazines and online forums, this was the appearance of "restored" cars that *weren't* pristine. "Yes, it is cool", a popular sticker associated with this trend read, and "No, I'm not going to paint it".[3] Unlike the "work-in-progress" cars of prior decades, these examples, complete with faded paint, patches of rust and worn and stained upholstery, were considered *finished*. Carefully rebuilt to run and drive like new, or often better than new, without obliterating the visible marks and scars of their many years of service, these "patina rides" quickly nudged their way onto the old-car scene. To be fair, outliers – historically-significant vehicles in very rough condition, as well as unrestored

---

2   A number of synonyms for "old-car hobbyist" are in common use, including "old-car gearheads", "restoration enthusiasts", "old-car restoration enthusiasts", "old-car enthusiasts", "classic-car enthusiasts", and the like. I use these terms interchangeably in this essay. On the demographics of the old-car hobby, see Lucsko, David N.: The Business of Speed: The Hot Rod Industry in America, 1915–1990, Baltimore: Johns Hopkins University Press 2008, p. 224–226; Id.: Junkyards, Gearheads, and Rust: Salvaging the Automotive Past, Baltimore: Johns Hopkins University Press 2016, chs. 2 and 4; Cross, Gary S.: Machines of Youth: America's Car Obsession, Chicago: University of Chicago Press 2018, ch. 8.

3   These have been available for years; see, for example, vintagevolks: advertisement, https://www.thesamba.com/vw/classifieds/detail.php?id=2103221 (accessed 13.08.2018).

"survivors" in good, if not pristine form – had long been part of the hobby.[4] What was genuinely new in the early twenty-first century was the *extent* to which patina had transfixed the old-car world.

This paper investigates this trend. Following a brief terminological discussion, it begins with a survey of the history of old-car restoration, focusing specifically on hobbyists.[5] Then, drawing heavily on broadly-circulated periodicals (including *Motor Trend*, *Cars and Parts*, *Hot Rod* and *Hot VWs*) and, for a more bottom-up angle on recent developments, important international forums like thesamba.com, it seeks out the origins of what I call the "patination turn" before ending with a discussion of its broader significance. For as we will see, beneath

---

4   On survivors and other outliers, see for example Dieter, R. E./Duerksen, Menno: "Tool Bag", in: Cars and Parts (C&P), Feb. 1972, p. 91; Unrestored Antique Auto Club of America: advertisement, in: C&P, Jun. 1974, p. 113; Bertilsson, Bo: "Super Type II x 2!", in: VW Trends (VWT), Jul. 1992, p. 78–79; Strohl, Daniel: "Historically Rich", in: Hemmings Classic Car, Dec. 2005, p. 58–63; RM Auctions: advertisement, in: Hemmings, Nov. 2007, p. 608; Anon.: "Wrecked Bugatti to Aid Charity", in: Detroit Free Press, 28 Jan. 2010, n. p.; Anon.: "HRG Returns to Family Fold", in: Classic and Sports Car (C&SC), Nov. 2010, p. 28; Anon.: "Best of 2010", in: C&SC, Feb. 2011, p. 12. There were also those among the old-car crowd who preferred rough-looking *and* rough-running cars, especially urbanites like *Car and Driver*'s Warren Weith; see for example Anon.: "Cars", in: Car and Driver (C&D), Mar. 1965, p. 20 and 22.

5   The literature on automotive restoration as a leisure-time activity is nearly nonexistent; exceptions include Lucsko, Junkyards, chs. 2 and 4, and Cross, Machines of Youth, ch. 8. The literature on automotive subcultures and enthusiasm more broadly is better developed; in addition to the titles cited above, see Dannefer, Dale: "Neither Socialization nor Recruitment: The Avocational Careers of Old-Car Enthusiasts", in: Social Forces 60 (1981), p. 395–413; Moorhouse, H. F.: Driving Ambitions: A Social Analysis of the American Hot Rod Enthusiasm, Manchester: Manchester University Press 1991; Post, Robert C.: High Performance: The Culture and Technology of Drag Racing, 1950–1990, Baltimore: Johns Hopkins University Press 1994; Bright, Brenda Jo: Mexican American Low Riders: An Anthropological Approach to Popular Culture, PhD thesis: Houston/Texas, Rice University 1994; Id.: Customized: Art Inspired by Hot Rods, Low Riders, and American Car Culture, New York: Harry N. Abrams 2000; DeWitt, John: Cool Cars, High Art: The Rise of Kustom Kulture, Jackson: University Press of Mississippi 2002; Lucsko, Business of Speed; Jakle, John A./Sculle, Keith A.: Motoring: The Highway Experience in America, Athens: University of Georgia Press 2008, p. 4 and 225–226; Kinney, Jeremy: "Racing on Runways: The Strategic Air Command and Sports Car Racing in the 1950s", in: ICON 19 (2013), p. 193–215.

its palpable sense of nostalgia and its outward obsession with the obsolete, the old-car hobby has actually long been predicated upon a very real reverence for novelty and progress. Only with the patination turn did a genuine sense of history – real, lived, weather-worn history – finally come to the fore.

REPAIR AND RESTORATION

What exactly do we mean by "restoration"? In his recent book *Together* (2012), Richard Sennett describes "restoration", "remediation" and "reconfiguration" as distinct forms of "repair". Restoration describes a repair in which one seeks to return a broken object to its original condition, ideally without leaving any traces of one's work behind. Remediation, on the other hand, is a form of repair in which one retains an object's original form and function while incorporating new or improved materials. Reconfiguration, the most radical of Sennett's three Rs, involves a thorough reconceptualisation of an object's form and function.[6]

Sennett's insights here are useful, and I will return to them later in this essay. But his terminology is slightly problematic, because in the automotive world both restoration and repair already carry meanings that are incongruent with his framework. To wit, when a part or system on a running car breaks down, you *repair* it.[7] When a more thoroughgoing overhaul to multiple systems is required, as is typical for a worn-out car or one that sat unused for a number of years, you *restore* it. A recent treatise on automotive preservation from the Fédération Internationale des Véhicules Anciens (FIVA, a global coalition of old-car clubs) seeks in part to clarify this distinction. Repair, according to its *Charter of Turin* (2013), is work done when a car is still in normal use, and "involves the adaptation, refurbishment or replacement of existing, damaged or missing components" in order to "make a vehicle fully operational again". This includes both "pragmatic repairs" (also known among old-car hobbyists as "bodge jobs": makeshift fixes, often using improvised materials, made to quickly return a vehicle to

---

6   Sennett, Richard: Together: The Rituals, Pleasures, and Politics of Cooperation, New Haven: Yale University Press 2012, p. 212–215; see also id.: The Craftsman, New Haven: Yale University Press 2008, p. 199–202.

7   On repair, see Borg, Kevin L.: Auto Mechanics: Technology and Expertise in Twentieth-Century America, Baltimore: Johns Hopkins University Press 2007; Pirsig, Robert M.: Zen and the Art of Motorcycle Maintenance: An Inquiry into Values, New York: William Morrow 1974; Harper, Douglas: Working Knowledge: Skill and Community in a Small Shop, Berkeley: University of California Press 1987.

service), as well as more careful "professional repairs" (the sorts of jobs done by skilled mechanics at a garage).[8] Restoration, on the other hand, involves the same kinds of tasks often seen in professional repairs but is done later, "with the aim of displaying the vehicle as it was at a particular point in time".[9] In other words, restoration brings together multiple kinds of repair work in order to return a vehicle to some prior state – typically, in practice, its state when showroom-new. Thus, while Sennett's conceptualisation of "restoration" as a form of "repair" makes some sense in theory, the terms are never used that way in practice. One would never call for help from the side of the road and tell the person who answers that one's car requires "restoration". That would be like going in for a routine dental cleaning and calling it "mandibular surgery". Likewise, one would never drag a beat-up car with a rusted-out body, shredded interior and seized mechanicals to a garage and simply say it needs "repairing". That would be like asking for a bit of floss to deal with a fractured jaw.

In sum, "repair", in real-world use, refers to the act of addressing a specific malady on a car, while "restoration" describes the much more comprehensive process of returning one to like-new form. Such has been the case for many decades. Well before the mass production of the Model T began, as Kevin Borg has shown, wealthy early adopters either worked on their cars themselves or hired chauffeurs both to drive them around and to care for their machines. And, as Borg's sources clearly show, the work that they performed was called "repair" (or, as a verb, "repairing" or "fixing"). Garages where owners stored their vehicles had "space devoted to repair of cars", workshops that catered to early motorists billed themselves as "repairing" centres, and schools offering training in all things automotive portrayed their programmes as "complete driving and repair course[s]".[10] Although the work was often more involved and much more

---

8   Fédération Internationale des Véhicules Anciens (FIVA): Charter of Turin, 2013, reprinted in FIVA: Charter of Turin Handbook, Brussels: FIVA 2017, p. 19–21, here p. 19.

9   FIVA, Charter of Turin, p. 19–21, here p. 20. Further discussion of the Charter of Turin appears later in this essay.

10 Borg, Auto Mechanics, quoted text from fig. 6, 7, 12 and 18. In addition to "repair", other terms crept into use over time – "maintenance", "lubrication", "service" – but these typically referred to various forms of preventive care, much more so than repair. For some examples, see Borg, Auto Mechanics, fig. 25: Witzel, Michael Karl: The American Gas Station: History and Folklore of the Gas Station in American Car Culture, Osceola, Wisconsin: Motorbooks International 1992: Jakle, John A./Sculle, Keith A.: The Gas Station in America, Baltimore: Johns Hopkins University Press 1994.

frequently required than would be the case in later years, it was certainly "re-pair" in the modern sense: discreet operations performed to address specific faults on otherwise roadworthy cars. Likewise, "restoration" has been the chief term used by old-car hobbyists to describe the process of returning a car to like-new form since at least the early 1950s, when it largely eclipsed the earlier and not entirely synonymous "refurbishment" (discussed in the subsequent sec-tion).[11] Stable across time, these common definitions make it difficult to lean on Sennett's terminology when studying the old-car hobby.

Further complicating matters is the fact that "restoration", in the automotive world, covers an array of activities mapping partly onto what Sennett calls "res-toration", and partly onto what he calls "remediation". This is due to the broad range of opinion, mentioned earlier, regarding the meaning of "like-new" – re-garding, that is, the precise ends towards which one should labour when engaged in automotive restoration. To those for whom the like-new standard means "per-fectly original" – i. e., precisely as a car looked and performed when it left the

---

11 Feature articles on the growing old-car hobby began to appear in the early 1950s (when the enthusiast magazine itself was still a new genre, having first appeared in the US in 1947 [see Lucsko, Junkyards, p. 249–250]). Among the first sixteen feature sto-ries and letters to the editor on the post-war hobby I have come across in my research thus far, all published between 1951 and 1957, "restore", "restored" or "restoration", deployed in the common-use manner I describe here, appeared in all of them, whereas "refurbish" or "refurbishment" are found in only three, the last of which appeared in 1954. On restoration, see Stratton, Charles L.: "Museum in the Rough", in: Motor-sport, Aug. 1952, p. 31 and 48–49; Gottlieb, Robert J.: "Classic Comments", in: Mo-tor Trend (MT), Oct. 1952, p. 36–37 and 55; Jaderquist, Eugene: "The Classic Thrill of Yesteryear", in: True's Automotive Yearbook (1952), p. 2–7 and 100–102; Gottlieb, Robert J.: "Classic Comments", in: MT, Feb. 1953, p. 54–57; Cetin, Frank: "He Brings 'Em Back to Life", in: Cars, Jun. 1953, p. 52–53 and 74; Gerrits, Russell: "Rod's Beginning", in: Rod and Custom (R&C), Jun. 1954, p. 58; Wherry, Joe H.: "Classic Cars from the Vintage Years", in: Car Life (CL), Sep. 1954, p. 52–55; Parks, George A.: "Perpetual Youngsters", in: CL, Dec. 1954, p. 42–45; Hegge, Robert: "Antiques in the Ozarks", in: CL, Mar. 1955, p. 55–58; Greenlee, Lyman: "So You Want to Restore a Classic, Part II", in: Motorsport, Jun.–Jul. 1955, p. 24–25, 46–48 and 50; Warren, Marian: "They Like 'Em Old", in: CL, Jul. 1955, p. 28–31; Anon.: "Restoring an Aristocrat", in: Motor Life (ML), Jan. 1956, p. 50–51; Anon.: "Long Live the A", in: Auto Craftsmen, Dec. 1957, p. 12–13. On restoration and refurbish-ment, see Schroeder, Bill: "Where the Ages Meet", in: Motorsport, Dec. 1951, p. 16–17 and 22–23; Wherry, Joe H.: "Model A Club", in: CL, Aug. 1954, p. 51–54; Bow-man, Hank Wieand: "Sutton's Shiny Scraps", in: Auto Age, Oct. 1954, p. 38–39 and 46. From the 1960s to the present, "restoration" has remained predominant.

factory – Sennett's "restoration" fits. But to those for whom it can mean "better than new" in some way – thicker sound deadening, improved rustproofing, or even engine and chassis modifications like electronic ignitions, smoother-riding shocks or better-quality tyres or brakes – "remediation" is a closer fit. Among gearheads, a number of more specific terms delineate the finer gradations: "concours restorations", "driver restorations", "sympathetic restorations", "resto mods", and so forth. But the broadest term of art across the hobby's many subgenres is simply "restoration".[12]

With all of this in mind, this essay uses the vocabulary of the old-car hobbyist: repair, restoration and (occasionally) refurbishment.[13] Synonyms, when they creep in, will be clear from the context.

OLD-CAR RESTORATION: A BRIEF SURVEY[14]

From the 1890s to the 1920s, automobiles were seldom "restored". At most they were "refurbished", typically to squeeze a few more years from them or to make them more attractive in the nascent resale market. Among owners, this could mean a thorough tune-up, perhaps some fresh paint and attention to the upholstery, and almost certainly a healthy dose of elbow grease to clean the whole car up.[15] Among dealers, who began to struggle with a second-hand surplus known

---

12 There are also many synonyms. "Fix" is often used for "repair", while "fix up" can mean either "restore" or "refurbish", depending on the context. In place of "restored" one will sometimes find "rejuvenated", "reworked", "rebuilt", "renovated", "redone" or, especially but not exclusively among non-native speakers, "renewed" (see Upstream Trading: advertisement, https://www.thesamba.com/vw/classifieds/detail.php?id=2172759 [from a Brazilian firm]; Porsche Kaiser: advertisement, https://www.thesamba.com/vw/classifieds/detail.php?id=2160769 [from a German individual]; ER Classics: advertisement, https://www.thesamba.com/vw/classifieds/detail.php?id=2154455 [from a Dutch firm], all accessed 14.08.2018. Also, there are a number of "re-" terms for the various tasks associated with a restoration, including "refinished", "repainted", "resprayed", "retrimmed", "reupholstered" and "rebodied".

13 The same applies to all of the specialist terminology I deploy in this essay: with rare (and clearly marked) exception, it is the language of the old-car hobbyists themselves.

14 This section draws heavily on my recent book: Lucsko, Junkyards, ch. 2.

15 Kilburn, Edwin: "Fitting a Mechanical Lubricator to an Old Car", in: Horseless Age (HA), 20 Mar. 1907, p. 399–400; Teachont, C.: "The Reader's Clearing House", in: Motor Age (MA), 4 Nov. 1915, p. 38; Anon.: "Editorial Perspectives", in: MA, 7 Mar. 1918, p. 10.

as the "used-car problem" in the 1910s, refurbishment nearly always involved similar measures, but could also entail far more radical procedures for improved marketability: fitting a new body to turn an old sedan into a speedster, converting a beat-up touring car into a truck, and the like.[16] Restoration, on the other hand, referred to the process of fixing up an old car not because it was otherwise unattractive or unserviceable, but instead because it possessed sentimental, historic or monetary value. Early on, restoration therefore stood apart from refurbishment less in process than in motivation. Unless an older car was modified extensively while being refurbished, that is, the process of restoring another would have looked much the same: a thorough clean, a tune-up, touch-ups to paint and upholstery, and the like.[17] But while one car might have been *refurbished* because its advanced age made it undesirable, another might have been *restored* because *its* age actually made it desirable.

At the dawn of the twentieth century, of course, "advanced age" meant something very different when applied to automobiles than it would a few decades later. The car itself was still a new technology at the time, and its fundamentals were evolving rapidly. Thus a three- or four-year-old car might indeed have seemed quite *old*, functionally and aesthetically. Cars like this, which the press called "ancients" or "relics", could be tough to resell. Some were refurbished by second-hand dealers and thereby found new homes, but often they met different fates. A few were scrapped, although this was uncommon at the time, while others, especially those still owned by wealthy early adopters who had since moved on to more up-to-date designs, were simply "tucked away in private

---

16 Anon.: "Remaking a 1910 Model", in: MA, 31 May 1917, p. 47; Barnes, Ray A.: "Comments and Queries", in: HA, 28 Jan. 1914, p. 164–165; Anon.: "Putting the Used Car Among Dealer's Assets", in: HA, 1 Nov. 1917, p. 18–19 and 88. See also Lucsko, Junkyards, ch. 1; Gelber, Steven M.: Horse Trading in the Age of Cars: Men in the Marketplace, Baltimore: Johns Hopkins University Press 2008, ch. 4.

17 Indeed, this is likely why the concept of "refurbishment" survived within the old-car hobby well into the 1950s (see above, footnotes 11–12), and even on occasion to the present: especially in the interwar and immediate post-war years, many of the old cars hobbyists worked on weren't yet quite so old, at least in terms of their condition, to require anything more than what a used-car lot might do to make an unappealing car less so. The same applies from time to time to modern hobbyists working either on the cars of the more recent past (the 1980s, for example) or on uncommonly well-preserved "survivors".

stables" and forgotten.[18] For in the 1890s and 1900s, sufficient time had not yet passed for these "ancients" to have become collectible, and thus restorable.

During the 1910s, however, elites on both sides of the pond began to pay attention to them. In 1911, two nobles sponsored an effort to track down early examples for display in a London museum.[19] In 1913, the organisers of a new-car expo in the US offered a prize to the oldest car that could make it to the show under its own power,[20] while in 1916 the Haynes Company launched an effort to locate the oldest of its cars still in service.[21] For its part, the automotive press in Europe and the US began to wax nostalgic about the early days of motoring through retrospectives and celebratory features on surviving "old timers".[22] Oblique references to do-it-yourself *restoration* began to crop up here and there in the 1910s and 1920s, too.[23] Bit by bit, an old-car hobby was emerging.

By the 1930s, the new pastime was sufficiently widespread among middle-class Americans[24] to support the founding of the Antique Automobile Club of America (AACA, 1935) and the Horseless Carriage Club of America (HCCA, 1937). Members of these organisations were chiefly interested in what they called "antiques", vehicles from the 1890s–1910s (HCCA) or the 1890s–1920s

---

18 Anon.: "What Becomes of Out of Date Cars?", in: HA, 31 May 1905, p. 600 (quote); Anon.: "What Becomes of Old Automobiles", in: MA, 23 Feb. 1928, p. 12; Anon.: "Where the Old Cars Go", in: The Automobile (TA), 20 May 1909, p. 841.

19 Anon.: "England Opens Its Motor Car Museum", in: MA, 27 Jun. 1912, p. 22; Fowler Dixon, W. S.: "John Bull Preserves Early Models of Power Propelled Vehicles By Placing Them in Motor Museum", in: MA, 2 Apr. 1914, p. 20–21.

20 Anon.: "Pennsylvania's 'Oldest Car'", in: HA, 28 Jan. 1914, p. 148.

21 Anon.: "From the Four Winds", in: MA, 8 Jun. 1916, p. 47; Anon.: "New Haynes for Old", in: MA, 19 Oct. 1916, p. 24.

22 Anon.: "Ancient History", in: TA, 23 Feb. 1911, p. 539, and 2 Mar. 1911, p. 627; Anon.: "Motoring of Early Days", in: MA, 6 Jun. 1918, p. 5–11; Duryea, Charles E.: "Motor Racing 20 Years Ago", in: MA, 27 May 1915, p. 24–28; Anon.: "The Oldest Car in Service", in: Motor, May 1911, p. 53; Anon.: "Priest Owns Oldest Car in the World", in: MA, 27 Jun. 1912, p. 22–23; Anon.: "10-Year-Old Cadillac Makes Fine Showing in English Test", in: TA, 6 Nov. 1913, p. 883; Anon.: "A Venerable Ride", in: Motor, Jul. 1913, p. 79–80; Anon.: "From the Four Winds", in: MA, 12 Aug. 1920, p. 56.

23 See for example Anon.: "Motor Junk Is Sold at Fair", in: MA, 3 Jun. 1915, p. 28; Anon.: "Reviving Old Model Car", in: MA, 27 Apr. 1922, p. 43.

24 My research to date on the period from the 1920s to the 1960s has focused almost exclusively on the restoration hobby in the US. Further work on the UK, continental Europe and Australasia in the coming years will allow me to flesh out the global scope of the early restoration hobby.

(AACA).[25] After the Second World War, the hobby broadened to include "classics", the luxurious and powerful makes of the 1920s–1930s: Duesenbergs, Mercedes, Rolls-Royces and the like. By the time old-car restoration hit its stride as a broad-based activity in the late 1950s, a third class had emerged. "Special-interest autos", as this niche was known, included the mass-market cars of the 1920s–1930s (mostly Fords, Chevrolets and Dodges).[26] Later, during the 1970s and 1980s, the special-interest class would expand to incorporate post-war mass-market vehicles, including a variety of imports.[27] Since then the hobby has grown a bit each year, as more and more cars "age in" (i. e. become "old" – old enough either to qualify for "historic" or "old-timer" status in their jurisdictions or club bylaws or to trigger a sense of nostalgia among those who owned or wished to own them when they were new).

OLD-CAR RESTORATION: METHODS AND MOTIVES

Every restoration is unique, because every vehicle endures its own mix of circumstances during and after its life on the road. Some have owners who meticulously maintain them, while others have less fastidious custodians. Some quickly rust from winter use on salted roads, while others begin to deteriorate only after they are retired from service (and left to the elements in storage lots, salvage yards or carports and other only half-protected spots). Others experience interior damage from UV exposure, suffer chassis fatigue from potholes and other on- and off-road hazards, or accumulate a patchwork of collision repairs. Some survive floods, others fires. A few endure most or all of these things. In short, by

---

25 Stubenrauch, Bob: The Fun of Old Cars: Collecting and Restoring Antique, Classic and Special Interest Automobiles, New York: Dodd, Mead, and Company 1967, p. 33.

26 These period definitions are based on Jaderquist, Classic Thrill, p. 2–7 and 100–102; Anon.: "How to Restore a $50.00 Classic", in: MT, Feb. 1953, p. 54–57; Wherry, "Classic Cars from the Vintage Years"; Gottlieb, Robert J.: "So Who Wants an Old Car?", in: MT, Nov. 1957, p. 42–43; Duerksen, Menno: "Free Wheeling", in: C&P, May 1969, p. 45–46; Cramer, Carl/Duerksen, Menno: "Tool Bag", in: C&P, Oct. 1969, p. 45; Switzer, Carol L./Duerksen, Menno: "Tool Bag", in: C&P, Sep. 1971, p. 68; see also Rae, John B.: The American Automobile: A Brief History, Chicago: University of Chicago Press 1965, p. 217–218. By the 1970s and 1980s, "classic car" had assumed a much broader meaning, see Lucsko, Junkyards, ch. 6.

27 Witness the advent of a *Hemmings* spin-off in 1970, *Special-Interest Autos*, which focused on the mass-market makes of the 1920s–1950s.

the time a car becomes a restoration project, it will have its own particular blend of mechanical, electrical, structural and cosmetic problems.

It can take many years to breathe new life into seized-up engines and corroded relays, and to put right mangled wiring, rotten upholstery and rusted-out bodies. Even matching the colour and sheen of factory paints can be a daunting task in practice, whether due to a dearth of original documentation or to the effects of various regulatory phase-outs of toxic substances over the last few decades.[28] Each of these types of restoration work deserves its own extended treatment, but for the sake of brevity I will focus here on body damage.

Wooden frames and body supports were used extensively on early cars, such that many antique and classic projects require the replacement of split or rotten components. By the 1970s reproduction wooden parts were available for common cars, like Model Ts. But right up to the present, replacing or repairing these elements has often required extensive fabrication, and thus woodworking skills.[29] For later all-steel vehicles, body problems come in a variety of forms, notably poorly-repaired collision damage and rust. Addressing both requires stripping the affected areas of paint, filler and active rust to reveal the extent of the damage. This is often done by hand, using elbow grease or an angle grinder, but for larger areas there have long been other options: sandblasting, various home- and shop-use chemical strippers and even full-body-shell acid dipping.[30]

---

28 On paint chemistry and the challenge of recreating colours and finishes, see Bock, Jürgen: "Handbook Paint Chapter", in: Charter of Turin Handbook, p. 95–115; Tutt, Gundula: Kutschenlack, Asphaltschwarz & Nitroglanz: Fahrzeuglackierung zwischen 1900 und 1945, Bielefeld: Delius Klasing 2018.

29 On reproduction of wooden elements, see Lester Groff: advertisement, in: Chevy Parts and Cars (CP&C), May–Jun. 1962, p. 26; Syverson Cabinet Company: advertisement, in: C&P, Apr. 1971, p. 78. On wooden fabrication, see Brown, Arch: "One Neat Pile of Kindling", in: C&P, Feb. 1983, p. 12; Anon.: "They Said It Couldn't Be Done", in: CP&C, May–Jun. 1962, p. 27; Anon.: "Rolls Draws the Wrong Crowd", in: C&SC, Sep. 2018, p. 37.

30 Lehman General Sales Co.: advertisement, in: C&P, Oct. 1969, p. 22; Tyrell, W. T.: advertisement, in: C&P, Apr. 1971, p. 48; Auto Strippers and Restorers: advertisement, in: C&P, Aug. 1974, p. 35; Smith, Everett F.: "Routing Rot", in: C&P, Jun. 1977, p. 156–159; Baskerville, Gray: "Scritch Takes It All Off", in: Hot Rod (HRM), Apr. 1975, p. 120, 122 and 124; Bryant, Thos L.: "Sports Car Restoration, Part 2", in: Road and Track (R&T), Jun. 1983, p. 154D, 156, 160, 162, 166 and 168; Kaho, Todd: "VW Rustoration, Part 2", in: HVW, Jun. 1991, p. 54–56 and 88. See also Erwin, Dan: "Project Junkyard Greyhound", in: European Car (EC), Apr. 2002, p. 102, 104 and 106–107, and Jun. 2002, p. 140 and 142–145.

Once a car is stripped and the scope of its problems is known, the real work can begin. Bent panels can be beaten back to form with hammers and bucks, while more extensively damaged (or rusted) metal often needs to be replaced. This can be done by drilling out the factory spot-welds, removing the offending panel and welding in a replacement by way of those same spot-weld points, resulting in a very clean repair. Just as often, however, the unsalvageable metal must instead be cut out mid-sheet and its replacement carefully seam-welded into place.[31] In more extreme cases involving damage to structural elements, temporary braces may need to be tacked in before the offending sections can be addressed safely. It all depends on where the damage happens to be.

Even when cutting and welding are not required, bodywork can be tricky. In theory, bolt-on panels like bonnets and wings can simply be replaced, but in practice the results can be subpar. Sometimes this is due to hidden damage, such as ever-so-slightly bent bodies to which new parts will not neatly fit. Sometimes it is due instead to poor-quality reproductions – bonnets with misaligned bolt-holes, say – or to the use of parts intended for models sold in other markets (which may look similar but have different mounting points or accessory cut-outs).[32] Because they tend to fit better, original panels sourced from salvage yards have long been favoured. But they haven't always been abundant, especially for antiques and classics. As *Rod and Custom* put it in 1954, describing the restoration of a 1910 Buick, "[y]ou can't [just] run down to the nearest wrecking yard and buy a fender or two".[33]

Why not? Many antiques and classics were never produced in large numbers to begin with, making the long-term survival of more than a handful of salvage-yard examples extremely unlikely. Also, many new-car dealers and manufacturers destroyed thousands of "old timers" during the interwar years to remove competing used cars from the market, and thousands more were scrapped for steel during the Second World War and the Korean War.[34] For the antique or

---

31 Anon.: "Dog Gone Rust", in: Super Chevy, Jan. 1997, p. 72–73, 75 and 76; Sly, James: "Restoration", in: EC, Dec. 2000, p. 136–145; Erwin, Dan: "Project Junkyard Greyhound", in: EC, Sep. 2002, p. 110 and 112–117.
32 Kirsten, Dean: "VW Fenders 101", in: HVW, Sep. 2002, p. 90–91 and 128.
33 Anon.: "The Past Meets the Present", in: R&C, Feb. 1954, p. 10.
34 1920s–1930s manufacturer/dealer scrappage: Lucsko, Junkyards, ch. 1; Gelber, Horse Trading, p. 71–72. Second World War: White, John R.: "Hemmings Offers Seed Money to Help Keep the Old-Car Hobby in Bloom", in: Boston Globe, 4 Sep. 1993, p. 45; Zimring, Carl A.: Cash for Your Trash: Scrap Recycling in America, New Brunswick: Rutgers University Press 2005, ch. 4. Korean War: Mullaney, Thomas E.: "U.S. to Step Up Auto-Graveyard Scrap Flow", in: New York Times, 4 Nov. 1951,

classic enthusiast of the 1950s, the relative dearth of salvage-yard spares left three options: hiring a specialist to fabricate parts; locating others with similar cars and swapping parts; or finding and buying a "parts car", a wrecked or hopelessly worn-out vehicle used as a rolling source of spares.[35] The first was out of the question for most budget-minded do-it-yourselfers, while the third grew tougher every year.[36] This left the second, and by the early 1960s, swap meets and classified-heavy periodicals like *Hemmings* (1955) and *Cars and Parts* (1957) had become vital pillars of the antique- and classic-restoration hobbies.

Others had better luck sourcing spares. This was especially true of the special-interest hobbyists who emerged in the 1950s. Because they favoured the mass-produced cars of the 1920s and 1930s – Fords, Chevrolets and the like, which were abundant in American salvage yards – they had little trouble finding what they needed.[37] This would remain true for many years, in part because of the sheer number of mass-market cars produced in the 1920s and 1930s, but also because the special-interest niche itself continued to expand. By the 1980s, it included everything from 1940s Fords and 1950s Cadillacs to 1960s muscle cars, and even many post-war imports. Salvaged parts for all of these were plentiful and cheap.[38]

Reproduction parts also trickled into circulation after the Second World War. No full study of the aftermarket restoration business has yet been done, but a preliminary survey suggests that by the early 1960s, a number of firms sold "repro" body panels, engine parts, upholstery, rubber, trim, tyres and other items for common special-interest cars.[39] Also, enthusiasts with less-common vehicles

---

p. 143; Anon.: "NPA to Order More Old Cars Scrapped", in: Automotive Industries (AI), 15 Dec. 1951, p. 20; Gottlieb, Robert J.: "Classic Comments", in: MT, Oct. 1953, p. 36–37 and 55.

35 Stratton, "Museum in the Rough", p. 49.

36 Parks, "Perpetual Youngsters", p. 42–45.

37 Wherry, "Model A Club", p. 51–54; Jaderquist, Gene: "Classics and Antiques", in: Motorsport, Nov.–Dec. 1954, p. 33–36; Anon.: "Long Live the A", p. 12–13.

38 Lucsko, Junkyards, chs. 2–3.

39 A few examples from among the many in evidence by the early 1960s: Ford Parts and Supply Company: advertisement, in: Antique Parts and Cars (AP&C), Jul. 1962, inside cover; Jim Sharman: advertisement, in: AP&C, Jul. 1962, back cover; Lester Groff advertisement, in: AP&C, Aug. 1962, inside cover; William Farhy: advertisement, in: AP&C, Aug. 1962, inside cover; and advertisements from Jim's Auto Parts, L. G. Carlson, Burchill Antique Auto Parts, Max Merritt, Robert E. Darr and Les Leather (all in: AP&C, Oct. 1963, p. 114). There is no need to further belabour the point: the reproduction-parts business was clearly on its way (though many of these

sometimes hired workshops to produce batches of scarce components, keeping what they needed and selling the rest to help underwrite the cost.[40] As the old-car hobby continued to expand, so did the repro trade. By the early 1980s, extensive portfolios were available for virtually any car one might hope to rebuild, from VWs and Fords to Packards and Bentleys. Even complete bodies were available for several makes – "the single most fantastic thing that has ever happened in the history of the universe", as one gearhead put it when new bodies for 1960s Camaros became available in 2004.[41] Not all of these aftermarket parts were necessarily high-quality: repro trim, rubber, and interior components, like the aforementioned poorly-fitting body panels, weren't always up to factory form, let alone the better-than-new standard to which many aspired.[42] Nevertheless, their broad availability made restoration simpler. After all, why bother repairing a cracked engine block, mangled wing or malfunctioning relay when one could just replace them and move on? Why indeed.[43]

---

simply bear the name of an individual, suggesting origins similar to those detailed in footnote 40).

40 A few examples: Greenland, Sheldon: advertisement, in: C&P, Oct. 1969, p. 30 ("exact reproductions" of antique brass and other trim parts); Hirsch, Bill: advertisement, in: C&P, Apr. 1971, p. 41 (a one-time deal for Packard and Continental hubcaps reproduced by the shop that made the originals years earlier); Sommer, Don: advertisement, in: C&P, Apr. 1971, p. 52 (reproduction 1930–1931 Chevrolet bonnet ornaments). As other cars joined the old-car hobby over time, early efforts to supply parts often came from enthusiasts as well. During the 1980s, for example, when VW Microbuses "aged in" *en masse*, enthusiasts like Keith Woerle of San Diego began to reproduce small parts such as camper roof-rack clamps (see "Transporter Talk", in: HVW, Oct. 1987, p. 16–17).

41 An early example was the all-steel pre-war roadster body produced by Experi-Metal Inc. of Michigan in the early 1980s (advertisement, in: Street Scene (SS), Sep. 1983, p. 17); by the 2000s, brand-new, all-steel bodies for 1950s Chevrolets, early Camaros and even MGs were available (the MG bodies being factory reproductions sold through MG Heritage). See Bowling, Brad: "Jet Age Revival", in: C&P, Jan. 2008, p. 16–19; Anon.: "Bench Racing", in: HRM, Dec. 2004, p. 18 (quote); http://www.bmh-ltd.com/mgbshell.htm (accessed 09.10.2018).

42 See Kimball, Rich: "Need an Original Interior for that '50s Beetle?", in: HVW, Jul. 1996, p. 70–71 and 127.

43 A few examples from earlier times, when many parts *had* to be rebuilt because replacements weren't available: Deeter, Robert L.: "The Tool Bag", in: C&P, Jun. 1977, p. 145 (cracked engine-block repair); Brown, Arch: "Maxwell Model '35'", in: C&P, Feb. 1983, p. 22–23 (fabrication of a replacement bonnet); Anon.: "High-Herm", in: HVW, Oct. 1987, p. 74–75 (fabricating replacement interior panels).

In time, this aftermarket abundance contributed to the rise of "over-restoration", for it allowed one to replace not just what was worn *out*, but also what was merely worn *in*. In other words, many finished projects looked and ran like new because they *were* substantially new. Like Sennett's craftsmen who "remediate", many also availed themselves of improved materials and methods as they strove for perfection – witness Ofner's cad-plated hardware and powder-coated chassis or any of countless others finished with multiple coats of hand-rubbed base- and clear-coat paint, treated with anti-rust compounds or fitted with better-quality wing beading, more robust upholstery or odometers rebuilt and re-set to zero.[44] Yet apart from the occasional complaint about overwrought repli-cas,[45] none of these practices stood out very much in the 1970s, 1980s or 1990s. They were perfectly in sync with the overarching "like-new" mantra.

True of those who aimed for factory originality as well as those who did not mind a tweak or two for drivability, durability or comfort, the old-car world's widespread fixation on *pristine newness* echoes, at least in part, an important as-pect of Sennett's craftsmanship: "the desire to do a job well for its own sake".[46] What counts as a job well done? Sennett explains that "[i]n the traditional world of the archaic potter or doctor, standards for good work were set by the commu-nity, as skills passed down from generation to generation".[47] Likewise, in the old-car world, the "showroom-new condition" standard was, for many decades and for several generations, simply what one meant by "job well done".

Why? Here one needs to read between the lines. Across the years, an open embrace of all things old has been widespread among restoration hobbyists: parts and supply houses with "obsolete" in their names, magazines with "old" in their titles, car shows at which various aspects of long-bygone pop culture are

---

44 See above, footnote 1; Alhadeff, Jere: "Full Circle", in: HVW, Nov. 1984, p. 63; Gregory, Fred M. H.: "The $64,00 Paint Job", in: C&D, Jul. 1993, p. 107–108, 110 and 113–114; Smith, Robert K.: "Great Lakes Ghia", in: HVW, Jul. 1996, p. 102; Id.: "Restored by Ruth", in: HVW, Dec. 2002, p. 66–69; Cooper Classics: advertisement, in: C&P, Oct. 1969, p. 7; Anon.: "New Products for Restorers", in: AP&C, Jul. 1962, p. 35. The use of better upholstery and reset odometers continues to the present; see for example Black Dog 1999: advertisement, https://www.thesamba.com/vw/classi fieds/detail.php?id=2161196, and CFRI 2899: advertisement, https://www.thesamba. com/vw/classifieds/detail.php?id =2154528 (both accessed 27.09.2018).

45 See Worthington-Williams, Mike: "Opinion", in: C&SC, May 1990, n. p.

46 Sennett, Craftsman, p. 9.

47 Ibid., p. 25.

carefully recreated through period props and costumes, and so forth.[48] But not until the recent patination turn did any of this actually mean "old things" as such. Instead it meant old things *when they were new*. It was a broad nostalgia for the moment an old car left the factory, or when it first hit the showroom, or when it was first parked in the driveway with pride – moments when it represented *progress*. In a sense, enthusiasts have long cherished old cars made to look and drive like new because they offer tangible connections to those moments. They bring back to life the progress of the past by showing off Ford's cutting edge in 1932, Packard's in 1955 or Porsche's in 1967.

As for all the years *since*? One's task, in the old-car hobby, was to make it seem as though they had never happened.

THE PATINATION TURN

Twenty-two years after Ofner's Beetle graced the pages of *Hot VWs*, a very different sort of car appeared in *Octane*. This too was an early Volkswagen, and it too had been painstakingly restored. The difference was that this vehicle, a 1952 van, was hardly better-than-new, or even like-new. It had no hubcaps, its seat was bound with duct tape, and most of its exterior was covered in rust. It looked as though no restoration work had ever been attempted, yet its owner, Ben Laughton of Dorset, UK, had worked for several years to get its look and feel just right. Discovered in a Swedish forest, where it had rested on its side for more than fifty years, the van was so fragile with rust that it proved impossible to pull it from the woods without further damaging its brittle bones. Instead, Laughton cut it into several smaller and much more stable sections, piled them on a trailer and headed out of the woods. Once home, he carefully stitched it

---

48 A few examples of "obsolete" parts houses across the years: the Obsolete Parts Company of Gallup, New Mexico (1970s); Parts Obsolete of Costa Mesa, California (1990s); BFY Obsolete of Orange, California (1990s); Obsolete Buick Parts of Dallas, Texas (2000s); and GM Obsolete of Phoenix, Arizona (2000s). Among periodicals, *Old Cars Weekly* (U.S.) and *Old Autos* (Canada) have been published for many years. Regarding pop culture, see for example Janes, J. George: "Antique Antics", in: Motorsport, Dec. 1954, p. 8–11 and 36; Coleman, Bill: "New England Street Rodders 10th Annual Swap Meet", in: SS, Mar. 1984, p. 30; Anon.: "Life in the Past Lane", in: HRM, Jan. 1990, p. 21; Iola Old Car Show: advertisement, in: C&P, Apr. 1992, p. 171; the outfits and atmosphere at the UK's annual Goodwood Revival and the US's annual Amelia Island Concours (https://www.goodwood.com/flagship-events/goodwood-revival and https://www.ameliaconcours.org, both accessed 05.10.2018).

back together. "Wherever possible", *Octane* explained, he "saved original panels, although they invariably needed fresh metal letting into the edges". And "[w]here there was simply not enough metal left to save", he "tried to use appropriate-age donor panels", seam-welded and carefully blended with the surrounding metal. After refinishing the chassis and fitting rebuilt running gear, the van was done. It looked like hell, but ran like new – exactly what he wanted.[49]

Laughton's van was *not* an unfinished project, nor was it a neglected clunker. It was also clearly not a well-preserved original of the sort one often sees at auction.[50] Instead, like many others owned by twenty-first-century gearheads – vehicles ranging from 1930s–1960s sedans and muscle cars to 1940s–1950s off-roaders and delivery vans, as well as countless air-cooled Volkswagens like his own[51] – Laughton's was a "patina ride". It was rebuilt not to erase but to *save* its marks and scars, the evidence of all that it had been through since it was new. This may seem reasonable, or even second nature, to anyone familiar with other kinds of collectibles. But in the context of the old-car world, a bit more work remains to better account for cars like this. Why, after seventy-plus years of chasing factory-fresh perfection, did some begin to value wear and tear? Why did glossy finishes and crisp upholstery start to lose ground to faded paint, surface rust and ragged seats? How might we make sense of the old-car hobby's patination turn?

Five related factors figure in. The first was the arrival, in the early 1990s, of "rat rods". Like the rough-and-tumble, race- and road-going hot rods of the 1930s – and decidedly unlike the pristine street rods of the 1970s and 1980s – rat rods, at the outset, were older cars modified for performance and drivability, rather than aesthetics. Also known at first as "beaters",[52] they were meant to be

---

49 Dixon, Matt/Laughton, Ben: "The Ultimate Barn (Door) Find", in: Octane, May 2018, p. 92–100, quote p. 99.

50 Anon.: "To Restore or Not to Restore? That Is the Question", in: R&T, Apr. 2007, p. 116. See also above, footnote 4.

51 Examples abound. For instance, Freiburger, David: "Love Your Rust", in: Car Craft (CC), Sep. 2005, p. 10; Smith, Jeff: "The $3,500 Challenge", in: CC, Dec. 2006, p. 54–59; Anon.: "Early Landies Go Home for 65th", in: C&SC, Sep. 2013, p. 21; Davis, Jacob: "Mother Nature's Masterpieces", in: HRM, Sep. 2018, p. 50–52, 54, and 56; Hunkins, Johnny: "The Survivor Phenomenon", in: CC, Dec. 2018, p. 4; Nalu33: advertisement, https://www.thesamba.com/vw/classifieds/detail.php?id=2176378 (accessed 09.08.2018); 57dbldoorpanel: advertisement, https://www.thesamba.com/vw/classifieds/detail.php?id=2203357; paul leahy: advertisement, https://www.thesamba.com/vw/classifieds/detail.php?id=2179512 (both accessed 20.08.2018).

52 See Freiburger, David: "Order Rodentia", in: HRM, Nov. 2010, here p. 45.

enjoyed, not coddled. Early rat rodders therefore spent the bulk of their time and energy under the bonnet, rather than on paint or upholstery. As a result, their finished projects tended to look rather "ratty", though this was mostly due to benign neglect and not design. Then, over time, this shabby aesthetic assumed a central role in its own right. Consequently, by the early 2000s many rat rodders were spending a considerable amount of time on their vehicles' appearance in a somewhat paradoxical effort, often involving the deliberate cultivation of faded paint and rust, to make it seem as if they did not care about the way they looked – or rather, more precisely, to make it *known* that they *enjoyed* their coarse and tarnished look.[53] And as this counter-aesthetic spread, it helped to pave the way for vehicles like Laughton's: rough and rusty, but ready to roll.

Ready to roll: therein lies the second factor. Simply put, part of the appeal of a patina ride is that it can be enjoyed in ways that other kinds of restored cars cannot. Consider "concours restorations". These are vehicles that have been rebuilt to uncompromising standards of perfection in every interior, exterior and mechanical detail so that they can be displayed at highly-competitive annual events like the Pebble Beach Concours d'Elegance in California or the Concours of Elegance at Hampton Court Palace in the UK. Also known derisively as "trailer queens", concours-level restorations are almost never actually driven. Instead they are carried from show to show in fully-enclosed transporters. "Driver restorations", on the other hand, are very nicely restored old cars that are meant to be enjoyed now and then. Those who own them are not concerned about maintaining concours levels of perfection, but they also aim to keep them as nice as possible. As a result, the occasional use a driver restoration tends to see is fairly gentle – sunny weekend cruising, trips to local shows and historic events and the like – rather than the grind and hazards of a daily commute. On the other hand, those who own old cars with "imperfections", everything from minor dings to heavy patination, can and often do enjoy them every day.[54]

The third factor was that from the 1950s to the 1980s, old-car hobbyists began to spend a lot of time in salvage yards. At first this was largely utilitarian: especially in the 1950s and 1960s, when the restoration aftermarket was in its infancy, boneyards were vital sources of parts. But over time, many gearheads developed an appreciation for salvage yards themselves, *and* for the rusty hulks within them, that went well beyond utility. Popular magazines played a visible role in this shift. Utilitarian parts-hunting features ran regularly in a number of periodicals in the 1950s and 1960s, but slightly more wistful stories about

---

53 On rat rods, see Lucsko, Junkyards, ch. 2, and Cross, Machines of Youth, ch. 8.
54 See for example Hunkins, Johnny: "Embrace Imperfection", in: CC, Jun. 2019, p. 4.

"junkyards we have known and loved" began to appear in the early 1970s and were common by the middle of the 1980s.[55] By far the most active in this regard was the restoration-oriented *Cars and Parts*, which ran monthly features on specific old-car yards for more than three decades beginning in the early 1980s. These photo-rich spreads quickly became reader favourites, and other publications across the globe were fast to follow suit. Part of the appeal lay in the details: readers relished the challenge of identifying the mangled cars that appeared and pointing out when and where the magazines themselves had misidentified them.[56]

Equally important, however, was the almost palpable sense of loss their junkyard coverage addressed. For by the 1980s, due to changes in environmental regulations, zoning laws, insurance requirements and the used-parts market, salvage yards with healthy stocks of old cars were dwindling in number. At the same time, prices for many vintage vehicles were rising to levels unsustainable for the average hobbyist, and safety and environmental regulations often made it difficult to own and drive an older car. With all of this in mind, those old-time wrecks that did remain in salvage yards provided a tangible and, for gearheads, much-needed connection to a golden age of automobility that seemed to be receding more and more each day.[57] "Their sun-faded colors contrast vividly", *Car and Driver* therefore noted in a wistful story about junked cars back in 1988, with "memories still bright in the mind. Each ... has a story to tell, and no way to tell it".[58] Ten years later, *VW Trends* noted in a similar feature that those who venture into salvage yards to see their wrecks up close "often have thoughts on

---

55 Early, utilitarian coverage: Francisco, Don: "A V-8 in Your Model A", in: HRM, Dec. 1951, p. 10–15, 43, 45, 47–48, 50–51 and 53–54; Phelps, Jack: "Bargains in Horsepower", in: ML, Nov. 1955, p. 26–27 and 66; Francisco, Don: "How to Buy Used Engines", in: CC, Feb. 1957, p. 12–17 and 58–59; Walordy, Alex: "Replacement Parts for Your Car", in: CL, Sep. 1958, p. 32–33 and 60–63. Later, more wistful coverage: Anon.: "Junkyards We Have Known and Loved", in: Special Interest Autos, May–Jun. 1971, p. 52–53; Hubit, Gregory: untitled letter, in: C&P, Apr. 1973, p. 114; Mayall, Joe: "Curbside", in: SS, May 1981, p. 6; Benty, Cam: "The Salvage Yard Maze", in: Popular Hot Rodding, Aug. 1982, p. 76–79; Anon.: "Happy Hunting Grounds", in: C&D, Jan. 1988, p. 86–87; Smith, Jeff: "Treasure of Sierra Vista", in: HRM, Jun. 1989, p. 135–137. See also Lucsko, Junkyards, chs. 2 and 4.

56 See for example Spitoleri, Tony: "Misidentified Junk!" in: C&P, Jan. 1982, p. 158; Bostick, Kenneth V.: "Salvage Yard Devotee", in: C&P, Nov. 1995, p. 167; Smith, Robert E.: "Salvage Yard ID", in: C&P, Aug. 1997, p. 9.

57 See Lucsko, Junkyards, chs. 3–6.

58 Anon.: "Happy Hunting Grounds", p. 86–87.

how 'life' may have been for [them] … 'Where have they been?' 'Who owned them?' and 'How did they get here?'"[59] By actively connecting their readers' collective nostalgia to the forlorn hulks they found in old-car yards, these stories helped them shed their distaste for the less-than-pristine.[60]

The fourth factor goes back a long way, is less direct and thus requires a bit more exposition. In 1931, a group of preservation experts convened in Greece as the First International Congress of Architects and Technicians of Historic Monuments, resulting in a statement of principles called the *Athens Charter*. Thirty-three years later, a second congress gathered in Italy to develop a stronger statement, the *Venice Charter*. Of the latter's sixteen detailed articles, the most germane to our discussion here are numbers 9 and 11. Article 9 declares that restoration work aims, above all, "to preserve and reveal … aesthetic and historic value". But, as Article 11 clarifies, this does not require a wholesale return to a pristine state. Instead, a proper restoration should preserve the layered history of its subject: "[t]he valid contribution of all periods … must be respected, since unity of style is not the aim."[61] However, not only does this stipulation evade the issue of historical authenticity, as Robert Russell has pointed out, by relativising periodisation; it also ignores the fact that "unity of style" is sometimes "precisely the aim of a particular restoration".[62] Still, the *Venice Charter*'s influence has been strong, especially among gallery and museum curators, for it gave them a useful set of principles for contemplating accuracy, restoration and over-restoration.[63] It also inspired those who work with several kinds of historic vehicles to draw up statements of their own: the *Barcelona Charter* of 2003 (ships), the *Riga Charter* of 2005 (rail) and, most recently, the *Charter of Turin* of 2013 (automobiles).

Like the *Venice Charter*, the *Charter of Turin* was the work of a group of experts in the field (automotive history and preservation, in this case) who wanted

---

59 Collazo, Frederick Inocencio: "Volkswagens in the Yard", in: VWT, Feb. 1998, p. 48.

60 Lucsko, Junkyards, chs. 2 and 4.

61 Second International Congress of Architects and Technicians of Historic Monuments, International Council on Monuments and Sites (ICMOS): International Charter for the Conservation and Restoration of Monuments and Sites (Venice Charter), Venice: ICMOS 1964.

62 Russell, Robert: "Abstraction, Authenticity, and the Abolition of Time", in: Hardy, Matthew (ed.): The Venice Charter Revisited: Modernism, Conservation, and Tradition in the 21st Century, Newcastle upon Tyne: Cambridge Scholars Publishing 2008, p. 99–106, here p. 102.

63 See for example Deck, Clara: "Conservation of Big Stuff at the Henry Ford – Past, Present, and Future", in: Objects Specialty Group Postprints 13 (2006), p. 168–183.

to develop guidelines for the treatment of authentic artefacts among practitioners (museum professionals and restoration specialists as well as ordinary enthusiasts). Much like the *Venice Charter*, it never quite nails down what it means by "authenticity". Instead it focuses on "historically coherent states" – a 1975 model-year car as it looked and drove in 1975, for example, or as it looked and drove ten years later, or twenty. Thus the key to a successful restoration – defined as an effort to return a vehicle to a chosen point in time – is preparatory research. Only by examining documents like build sheets, photographs and receipts, and by carefully collecting clues from the vehicle itself, can one be certain of the finished product's historical coherence. Otherwise, one risks mixing periods and ending up with a car in a "disrupted state". Tellingly, the *Charter* lists among disrupted-state examples those that are "intentionally shown in a superficially neglected and damaged condition on the surface" but which mechanically "have been restored to a good standard or even renovated and tuned".[64] This fits vehicles like Laughton's well. If taken out of context, though, it does appear to preclude most patina rides from serious consideration among the cognoscenti – after all, they are "disrupted", not "restored". But other sections of the *Charter* make it clear that vehicles with heavy patination are in fact quite welcome, much more so than over-restored examples "which exaggerate an imaginary mint condition" by "extinguish[ing] every 'annoying' or 'unsightly' trace of age and … historic substance."[65]

To bring us back to the matter at hand, factor four in the old-car hobby's patination turn might best be summed up as a broader shift in attitudes about material authenticity. And although I have thus far found but a single reference to the *Venice Charter* in the vast array of literature on the old-car hobby,[66] the fact that

---

64 Charter of Turin, p. 20–21 (all quoted material in this section appears on these pages).

65 Kohler, Thomas: "Authenticity and Authentic Restoration", in: Charter of Turin Handbook, p. 47 (quote). Passages more friendly to the preservation of patina in the *Charter* and its accompanying essays can be found on p. 16, 19, 20 and 99–102.

66 Though still incomplete, my survey of the literature has thus far only turned up Strohl, Daniel: "FIVA Enacts Charter of Turin, Establishes Historic Automobiles as Cultural Artifacts", in: Hemmings Classic Car, 14 Feb. 2013, https://www.hemmings.com/blog/2013/02/14/fiva-enacts-charter-of-turin-establishes-historic-automobiles-as-cultural-artifacts (accessed 17.01.2019). The following also warrants note: enthusiast periodicals frequently publish stories on specific museums, but they almost never say anything at all about the preservation practices those museums use. And when the subject of preservation in the museum world does come up, it can be quite contentious; see Hopkins, Kenneth W.: untitled letter, in: C&P, Sep. 1970, p. 80.

it helped inspire the *Charter of Turin*[67] does at least confirm that it has played a role in this broad shift.

So too has a fifth and final factor: the growing importance of originality within the old-car hobby. For many years, the pursuit of like-new perfection has gone hand-in-hand with the pursuit of factory-correct details. Hence the hobby's passion for nicely-preserved, all-original vehicles, as well as for what gearheads call "new-old-stock" (NOS) components: original replacements that went unsold at the time. Such was the allure of these elusive parts that many spent as much time hunting for them as they did actually working on their cars.[68] Over time, many came to value the authenticity conferred upon their restorations by the use of NOS parts to such an extent that they began to actively seek out genuine *used* parts, too, even if they showed some signs of wear. They also started thinking much more carefully before replacing imperfect yet authentic and fully service-able components. In 1988, for example, long before the first patina ride as such was built, *Road and Track*'s Peter Egan found himself in an unexpected conun-drum while contemplating the original aluminium panels on his Lotus Seven project. "The side[s] … were slightly creased where someone had bent them back out of the way to reweld the frame", he explained, and "[t]here was a small dent in the flat rear-body panel where the nose of another car had no doubt leaned … to lighten its rear traction in a corner". Nevertheless, Egan decided to keep the racing-scarred originals rather than replace them with unblemished re-pros. Other fans of vintage sports cars helped him make the call. "'I never re-place anything I can save'", one Jaguar enthusiast explained. "'You see those old factory inspector's chalk marks on the back of a dash panel and you realize the whole car is full of English ghosts. If you let them escape, they never come back.'" So Egan kept the panels, warts and all.[69] But the resulting car was *not* a patina ride: he carefully restored the rest of the Lotus and ended up with a very presentable, well-preserved original with its history intact.[70]

In time, as others worked to save *their* projects' scars, the fever slowly spread, as did the number of rough but serviceable components left "as found". Delicate techniques for preserving (and sometimes faking) patination spread as

---

67 Charter of Turin, p. 7.
68 Tales of NOS "treasure hunts" began to appear in print at least as far back as the early 1970s, if not earlier. See Lucsko, Junkyards, ch. 4. On authenticity, see Orvell, Miles: The Real Thing: Imitation and Authenticity in American Culture, 1880–1940, Chapel Hill: University of North Carolina Press 1989, and Russell, "Abstraction, Authentici-ty, and the Abolition of Time".
69 Egan, Peter: "Side Glances", in: R&T, Nov. 1988, p. 24 (all quotes).
70 Ibid.

well.[71] In the 1990s and early 2000s, this was most clearly evident among those who built rat rods, for whom the warts themselves were often the entire point. But even hard-core rat-rod fans were well aware that something more was at stake in the decision to preserve a pile of patinated parts than simply finding out how far their "rougher-the-better" mantra could be pushed. At stake, to put it most succinctly, were those tangible connections to the past – those ghosts to which Egan alluded. As *Hot Rod*'s David Freiburger put it in 2007, "patina lends proof of life. It tells a saturated story of age, of history absolutely unrevised".[72] And this, perhaps more than anything else, was why rough-hewn rides like Laughton's van would soon be all the rage. After all, as many an old-car nut has muttered over the years, "it's only original once".[73]

THE PARADOX OF ORIGINALITY: RESTORATION AS INTERPRETATION

Or is it? Is a car with painstakingly-preserved patina truly "original" – or at least, any more so than one carefully restored to factory-fresh form? More to the point, can faded paint and worn upholstery really tell a "saturated story" of "history absolutely unrevised?"

On these and other crucial questions, many disagreed with the likes of Egan, Freiburger and Laughton. "[T]urned off by excesses in over-restoration", as *Road and Track*'s Matt DeLorenzo countered in 2009, the "'preservation' … camp has pushed the pendulum just as far in the opposite direction". "[I]t's beyond me," he continued, "how people can ooh and aah over an all-original wreck. This isn't so much preservation as it is neglect".[74] In the high-end world

---

71 Elliott, Kev: "Black and Chrome", in: R&C, May 2013, p. 14; Bortles, JoAnn: "Aged to Perfection", in: CC, Feb. 2019, p. 26–29; Hunkins, "Embrace Imperfection".

72 Freiburger, David: "Patina", in: HRM, April 2007, p. 61.

73 This common saying eventually became the title for a popular-market book: Lentinello, Richard: It's Only Original Once: Unrestored Classic Cars, Minneapolis, Minnesota: MBI Publishing 2008. For an example of the term in use, see "Patina", forum thread, https://www.thesamba.com/vw/forum/viewtopic.php?t=104480 &postorder=asc (accessed 03.06.2019).

74 DeLorenzo, Matt: "The Road Ahead", in: R&T, Dec. 2009, p. 11.

of auction houses, museums and concours d'elegance, many share his qualms, and the question of whether to preserve or restore remains hotly contested.[75]

So too in the broader world of do-it-yourselfers. "[Y]ou can restore a car hundreds of times but you can never unrestore it", one gearhead posted to a well-trafficked vintage Volkswagen forum, thesamba.com, in 2005. "Once you paint it the first time, it can't be undone." In a play on MasterCard's famous commercials from the 1990s and early 2000s, another poster agreed: "one week, 3 coats base, two clear, $5k. 40 years of hard work, personality, and character … priceless".[76] Several years later, others on that same site clearly saw things differently. "If I hear another 'they are only original once' justifying a rusted out piece of junk, I will puke", wrote one in 2011. "Bringing such a car back to original via restoration does far more to honor that car", s/he added, because "[a] broke down rust bucket is not 'original', it is just a neglected piece of junk that has nothing to do with how that car ran and looked originally". Another put it more succinctly. "Actually, semantically, the term 'Original Condition' is somewhat misleading: these cars did not roll out from the factory worn out!"[77]

Therein lies what we might call the paradox of originality. For while a patinated car may not *look* the way it did when it was new, it is certainly more "original" than a comprehensively restored example, at least insofar as its panels, paint and upholstery – however rusty, faded and worn – are likely those it came with on day one. At the same time, a restored car may *look* exactly as it did when it was new, but it is certainly less "original" than a patinated example because many of its parts and all of its paint and upholstery will have been replaced. Perhaps the best way out of this puzzle lies in recognising that "originality" is truly fleeting. In other words, "it's only original once" may well mean exactly that: *only once*, on day one.

Importantly, this paradox of originality extends to what comes *after* day one, too. For while it is surely true that an old and patinated car reveals its history in ways that others cannot, the things that it makes known are certainly not "absolutely unrevised". Consider Laughton's van. A tremendous amount of work was involved in rendering it ready for the road – arguably even more than typically

75 See for example Craft, B.: "Roddin' at Random", in: HRM, Jun. 2011, p. 20; Bremner, Richard: "Left Well Alone, or Barn Again?", in: C&SC, Aug. 2011, p. 110–113 and 115; Anon.: "To Restore or Not to Restore".

76 "Patina", forum thread, see footnote 73.

77 "What's Worth More: Restore or Survivor? Here's one Answer", forum thread, https://www.thesamba.com/vw/forum/viewtopic.php?t=582701&postorder=asc (accessed 03.06.2019).

goes into a concours-level restoration. And in performing this necessary work on his VW, Laughton and those who helped him did in fact revise its history. So too with nearly all of the patina rides in service: putting their histories on display requires a "disruption", in the words of the *Charter of Turin*, or the remediative intervention of a craftsman, in the words of Richard Sennett. Or, perhaps more accurately, it requires an approach to the material reality of a given car not unlike that of an historian working with a body of primary documents. Like any good work of history, it requires synthesis and interpretation.

And, like any good work of history, it is also far more apt than not to prompt a lively debate. This the patination turn has surely done.

OBSOLESCENCE AND DISPOSAL

Mending or Ending?

Consumer Durables, Obsolescence and Practices of Reuse,
Repair and Disposal in West Germany (1960s–1980s)

Heike Weber

Over the course of the 20th century, consumer practices of care, repair and disposal underwent substantial changes. "Ending is better than mending", whis-
pered a nightly recorded voice into the ears of those sleeping in Aldous Huxley's
dystopia *Brave New World* (1932) to encourage them to throw old possessions
away and buy new ones.[1] Two to three generations later, the indoctrination
seems to be routine for many consumers in the Global North. A recent survey
among Germans aged between 18 and 39 reported that half of them have never
had a pair of shoes repaired.[2] By contrast, at the time Huxley was writing,
around 40% of what Germans spent on shoes flowed into repair services.[3]

When consumers decide whether to "mend or end" a commodity, they decide
on the temporality of its use. But consumer product lifetimes have also changed
over time. When the term "consumer durables" – *langlebige Gebrauchsgüter* in
German – spread in post-war economic thinking, these items were less "durable"
than might be believed, and a second term soon followed: product "obsolescence", which expressed the notion that consumer durables "age" and users
might see them as "obsolete" when compared to the latest offerings. Whether a

---

1   Huxley, Aldous: Brave New World, Harlow: Longman 2008 (first published in 1932),
    p. 40.
2   Anon.: Wegwerfware Kleidung, Repräsentative Greenpeace-Umfrage zu Kaufver-
    halten, Tragedauer und der Entsorgung von Mode, Hamburg 2015, p. 73.
3   Sudrow, Anne: "Reparieren im Wandel der Konsumregime. Bekleidung und Schuhe
    in Deutschland und Großbritannien während des Zweiten Weltkriegs", in: Technik-
    geschichte 79, 3 (2012), p. 227–254, here p. 232.

consumer durable's lifetime is prolonged or terminated is thus co-determined by norms and values on temporal agendas such as novelty, transience and deterioration, alongside broader economic and sociocultural considerations. In the second half of the 20th century, mass consumers in affluent Western societies, for instance, acquired the latest car or television model also as a marker of status or lifestyle, and self-repair was more of a recreational activity than an economic need. By contrast, in socialist societies, repairing, reusing or hording items for later use or barter were personal strategies to overcome the shortcomings of the socialist market.[4]

In the United States, Susan Strasser has described how, after the nation transitioned into a mass consumer society in the interwar years, the traditional stewardship of objects decreased and throwing away became the norm for worn-out clothes or broken things; convenience and disposable products were now marketed on a mass scale and urban authorities installed waste services as public disposal channels.[5] But even given the disposability of goods and the availability of waste infrastructures, discarding still poses the problem for consumers of parting with goods – a problem for which terms such as "dispossession" and "divestment" have been introduced in material culture studies; as well as the issue of waste, it raises ethical questions of morals and norms, of values, meanings and their re-definition.[6] While the transitions of the United States and various Western European countries into mass consumption societies have been extensively described,[7] there is a lack of relevant research on using and discarding consumer durables: the issues of wear and tear, repair and the practices of care,

---

4   See Gerasimova, Ekaterina /Chuikina, Sofia: "The Repair Society", in: Russian Studies in History 48 (2009), p. 58–74; Möser, Kurt: "Thesen zum Pflegen und Reparieren in den Automobilkulturen am Beispiel der DDR", in: Technikgeschichte 79, 3 (2012), p. 207–226.

5   Strasser, Susan: Waste and Want. A Social History of Trash, New York: Metropolitan 1999.

6   Lucas, Gavin: "Disposability and Dispossession in the Twentieth Century", in: Journal of Material Culture 7 (2002), p. 5–22; Gregson, Nicky/Metcalfe, Alan/Crewe, Louise: "Moving Things Along. The Conduits and Practices of Household Divestment", in: Transactions Institute British Geographers 32 (2007), p. 187–200; Gregson, Nicky/Crewe, Louise: Second-Hand Cultures, Oxford/New York: Berg 2003.

7   For West Germany see: König, Wolfgang: Geschichte der Konsumgesellschaft, Stuttgart: Franz Steiner 2000; Haupt, Heinz-Gerhard (ed.): Die Konsumgesellschaft in Deutschland, 1890–1990. Ein Handbuch, Frankfurt a. M.: Campus 2009; Andersen, Arne: Der Traum vom guten Leben. Alltags- und Konsumgeschichte vom Wirtschaftswunder bis heute, Frankfurt a. M.: Campus 1999.

disposal, reuse and resale, once the affluence of the latter 20th century saw households acquiring more and more items.[8] Even on the quantitative level, detailed figures on equipment acquisition are absent, as are approximate numbers on the continuous growth in household possessions over time (while household sizes shrank) or on second-hand markets and the importance of private or professional repair and resale.

This chapter considers the case of West Germany between the 1960s and the 1980s, when consumers adopted novel values and ways of handling domestic equipment. From the 1960s onwards, post-war economic prosperity saw "citizen-consumers" – as historians would later term them[9] – shopping in supermarkets and acquiring novel consumer goods such as pre-packaged food, plasticware and mass-produced furniture. Items such as furniture and washing machines were now acquired without the idea of a lifelong or even intergenerational use in mind, and practices of care, reuse and disposal changed accordingly. Such everyday practices have left behind virtually no manifest historical sources, so the chapter approaches changing consumer culture indirectly, through three closely interlinked fields, namely discarding bulky waste, the conditions and channels for private and professional reuse, resale and repair, and, on the discursive level, debates on society's wastefulness and product obsolescence. Sources that document these fields include expert journals for waste practitioners, archival documents on waste collections, reports on or price tables from second-hand markets, surveys within the electrical profession, consumer journals, housekeeping or repair guidebooks, trade address books, popular and professional literature and the press.

The chapter begins by looking at the phenomenon of bulky waste (*Sperrmüll* in German). Emerging in the 1950s, this specific waste fraction represented a material trace of changing practices in acquiring and discarding household effects. Many consumers "ended" the use of consumer durables by discarding them on bulky waste heaps. The experience of local authorities in dealing with

---

8　Rough sketches include: Slade, Giles: Made to Break. Technology and Obsolescence in America, Cambridge, MA: Harvard University Press 2006; Trentmann, Frank: Empire of Things. How We Became a World of Consumers, from the Fifteenth Century to the Twenty-first, London: Allen Lane 2016, p. 622–755; König, Wolfgang: Geschichte der Wegwerfgesellschaft. Die Kehrseite des Konsums, Stuttgart: Franz Steiner 2019; Heßler, Martina: "Wegwerfen. Zum Wandel des Umgangs mit Dingen", in: Zeitschrift für Erziehungswissenschaft 16, 2 (2013), p. 253–266.

9　Cohen, Lizabeth: A Consumer's Republic. The Politics of Mass Consumption in Postwar America, New York: Knopf 2003; Prinz, Michael: "Bürgerrecht Konsum", in: Archiv für Sozialgeschichte 44 (2004), p. 678–690.

bulky waste and its removal gives us a glimpse into practices of discarding and the channels for disposal. Given households' increasing acquisitions on a quantitative level, discarding via bulky waste collections soon began to take precedence over handing down or reselling. However, even as bulky waste, these items had an "afterlife":[10] some were salvaged, repaired and reused, while most bulky waste was dumped in landfills whose specific combinations of materials eventually generated unforeseen long-term problems.

The following two sections are devoted to the channels and conditions of re-use, resale and repair, describing them as interlocking strategies which were co-determined by consumer offers and discarding practices. The first part provides an overview of repair and resale, mostly on the basis of professional markets or services; some sources also shed light on private reselling, handing down or self-repair practices. The second part looks at professional repair in the case of televisions and radios and shows that repair adhered to various economic developments depending on respective product categories. Services for advice, maintenance and repair were key for the widespread adoption of any sophisticated technical appliance,[11] while self-repair was restricted to expert users or, in the case of lay users, to simple fixes. As we will see, televisions were notorious for regular defects over several decades and their use relied on a booming field of professional repair. In contrast, post-war radio designs became less repairable, leading to a decline in the radio repair sector.

The chapter concludes with the post-war intellectual critique of the contentious notion of "throwaway society" (*Wegwerf-Gesellschaft*). This term was also prominent in a fierce dispute on product obsolescence which emerged in the 1970s. The debate grappled discursively with disposability, excess waste and the contemporary observation of a demise of repair and repairability as well as foreshortened product lifespans. But in the end, the discussions did virtually nothing to promote repairable or more durable designs; they failed to shed light on why and how users "ended" a consumer durable or to clarify how producers actually managed product lifespans.

The focus areas identified above – discarding bulky waste, the paradigms of repair, reuse and resale, and debates on wastefulness and obsolescence – were part and parcel of changing cultures of "ending" and "mending". The chapter carves out their historical situatedness in the context of larger social, economic

---

10 See Dhawan, Ayushi: "The Persistence of SS France. Her Unmaking at the Alang Shipbreaking Yard in India" (this volume).

11 Krebs, Stefan/Weber, Heike: "Rethinking the History of Repair: Repair Cultures and the 'Lifespan' of Things" (this volume).

and cultural changes: authorities redefined the households' problem of how to discard bulky items as a public obligation; the average affluence of households increased, as did their acquisitions; producers amplified their product offerings while innovation rates likewise accelerated; and cost ratios shifted in favour of accessing new equipment since rising wages hampered the labour-intensive reuse and repair markets.

"ENDING" CONSUMER DURABLES IN THE AFFLUENT AGE: BULKY WASTE AND ITS DISPOSAL

Like many other West German cities, Frankfurt am Main established a municipal disposal service for bulky items in the early 1960s. As Frankfurt's bulky waste service operated on call and involved a charge – disposing of a sofa cost 2 Deutsche Mark (DM) and a chair 1 DM –, staff documented any incoming item for whose final disposal the municipal waste service would be responsible:[12] over a period of five weeks in the spring of 1961, 125 couches, 5 chaise longues, 40 armchairs, 75 mattresses, 27 beds and three fridges were collected, along with some further small household items. In 1967, the yearly list read as follows:[13] 2,009 sofas and couches (i.e. more than 160 per month), 3,532 cardboard boxes, 2,190 boxes and suitcases, 1,011 spring frames, 2,135 mattresses and more than a thousand carpets, chairs, beds, armchairs and cupboards, 1,376 stoves, around a thousand shelves, 632 refrigerators and washing machines, and considerable numbers of bicycles, car tyres and other items.

As an unintentional historical record, these descriptions indicate the onset of households' growing need to dispose of old things once the early post-war period of thrift and make-do gave way to affluence and the increasing consumption described in consumption history.[14] In the course of two to three decades, equipment in kitchens and other rooms in the house changed as much as the channels and methods of discarding and disposal. By 1970, mass consumption had profoundly transformed Western European societies. Convenient heating

---

12 See the archives of the Institut für Stadtgeschichte (IFS), Magistrat an Stadtverordneten-Versammlung: "Müllbeseitigung und Straßenreinigung", 19 Jun. 1961, Stadtkämmerei, 2.060; on fees: Anon.: "Stadtreinigung holt sperrige Güter ab", in: Frankfurter Hausfrauen-Zeitung, Jul. 1965, p. 4.

13 IFS, Magistrat an Stadtverordneten-Versammlung: "Kehrichtabfuhr im Allgemeinen", 26 Feb. 1968, Magistratsakten, 6.961.

14 See König, Geschichte; Haupt, Die Konsumgesellschaft; Andersen, Der Traum.

systems were gradually replacing traditional fireplaces, which accounts for the more than thousand stoves in Frankfurt's bulky waste service in 1967. Mass motorisation had set in, and rural citizens were also becoming mass consumers. Moreover, consumers replaced appliances or furniture more often than the previous generation. By 1970, purchases of key appliances such as washing machines and televisions mostly served to replace older models, while the use of innovations such as cassette recorders, video recorders and dishwashers grew steadily. The share of possessions that were repaired or reused fell, as shown in the subsequent section on "mending".

Bulky waste (*Sperrmüll*) appeared as urban litter in the late 1950s, when some consumers discarded unwanted items by dumping them on street corners or in car parks, or driving to the outskirts of the town or city and dumping them there.[15] Small traders and warehouses also discarded packaging, and in the 1960s, abandoned cars began to appear. At first, municipal authorities were undecided as to whether these residues constituted a municipal responsibility.[16] But in their effort to keep towns and cities clean, many swiftly took on the task. By 1962, more than a hundred towns and cities with over 10,000 inhabitants were operating bulky waste services, according to an initial survey by the Association of German Cities.[17] In 1964, a national study on bulky waste was conducted, with the results affirming that urban households had turned into "considerable producers of bulky waste", with old mattresses, chairs, pianos, furniture, carpets, sewing machines and bicycles figuring as prominent items.[18] Some local representatives even called for municipal bulky waste services for towns with fewer than 10,000 inhabitants.[19] While "ending" cars was left to private dismantling channels and junkyards, bulky waste systems became a seemingly indispensable extension of municipal waste disposal services, with just one basic rule: any item, except for motor vehicles or rubble, was considered as potential bulky waste as long as it did not fit into municipal waste bins for disposal. While citi-

---

15 Mahlke, H.: "Sperrmüllabfuhr in Bochum", Städtetag 7 (1961), p. 395–397; Borchert, Fritz: Gutachten über die Sammlung, Aufbereitung und Beseitigung von Sperrmüll. Berlin/Munich: Schmidt 1964.

16 Langer, W.: "Geordnete Müllablagerung", in: Städtetag 1 (1965), p. 41–48.

17 Only 128 towns and cities replied; seven of them lacked separate bulky waste services. See Erbel, Alfons: Sperrmüllabfuhr. Ergebnis einer vom Deutschen Städtetag im Einvernehmen mit dem Verband Kommunaler Fuhrparks- und Stadtreinigungsbetriebe im Jahre 1962 durchgeführten Umfrage, Cologne: Deutscher Städtetag 1963.

18 Borchert, Gutachten, p. 4.

19 Magistrat: Kehrichtabfuhr: Auszug aus den Mitteilungen der Kommunalen Gemeinschaftsstelle für Verwaltungsvereinfachung, 25 Aug. 1963, Nr. 148/63.

zens had once been responsible for the disposal of their effects, now neither individual consumers nor producers but rather local authorities took over the duty
and logistics of durables' final disposal.[20] In rural areas without a waste service,
a gravel pit often served as an informal dumping ground.

Bulky waste emerged for several reasons. The 1964 bulky waste study pointed to rising incomes which enabled the purchase and replacement of furniture
and domestic appliances and the shrinking scrap, second-hand and repair
trades.[21] But alongside these economic reasons, social, cultural, material and
spatial aspects were decisive. Urban apartments and post-war housing lacked adequate storage space for people to hold on to items that were no longer being
used. Domestic fireplaces, once ignited by scrap cardboard or wood, disappeared, and novel consumer goods such as plastic bowls or pocket radios would
rarely be reused or repaired. Moreover, the traditional care or repair of furniture
and other possessions and practices of handing down or reselling old things were
substituted by a consumer culture dominated by a passing use of more and more
objects.

Popular books and magazines on housekeeping, overtly aimed at housewives, offer some insight into these changing disposal practices. Around 1960,
advice on reuse, e.g. how to make a sideboard from the old living room cabinet,
was still abundant.[22] But housewives were also encouraged to discard old items
to keep order. In particular, moving house was the perfect opportunity to "part
with all useless clutter" as described in a women's magazine: housewives should
sort through wardrobes, drawers, cellars and attics and get rid of the "gimcrack"
hoarded in them.[23] This article advised donating old clothes and furniture to
charitable organisations – channels for reuse which were still relevant in later
times but were forced into a niche when compared with the volumes disposed of
in bulky waste collections.

In the 1960s and 1970s, bulky waste services were available in more and
more urban regions and easy to use – they were free of charge or required a minimal additional fee. This contributed to the steady increase in the amounts collected. Volumes of bulky waste fluctuated between regions, but they were clearly related to collection arrangements and the sizes of municipal waste bins.

---

20 Strasser, Waste; Weber, Heike: Reste und Recycling bis zur "grünen Wende" – Eine
    Stoff- und Wissensgeschichte alltäglicher Abfälle, Göttingen: Vandenhoeck & Rup-
    recht 2021 (forthcoming).
21 Borchert, Gutachten, p. 4.
22 See e. g.: Anon.: "Was tun mit dem alten Schrank?", in: Brigitte 1 (1960), p. 67.
23 Anon.: "Wir ziehen um!", in: Brigitte 17 (1959), p. 43.

Towns and cities with public open-street collections had higher bulky waste volumes – often also including bags filled with textiles or paper – than those with fee-based pick-up services, an experience which led West Berlin to switch from street collection to a pick-up service in 1975. In regions lacking large municipal waste containers, bulky waste amounts reached higher levels. Figures on total amounts diverge, but they suggest that by the early 1980s, the average inhabitant – in urban as well as rural households – generated between 22.7kg and 35.6kg of bulky waste each year.[24] Detailed studies on material contents are rare, but an analysis for the mid-1970s in the district of Ludwigsburg reported that paper and wood represented more than half the weight of waste, followed by paperboard (16 to 19%), plastics (around 7%) and iron (5 to 8%); glass and textiles contributed around 3 to 6%.[25] By the late 1970s, more and more electrical consumer appliances were also being found in the content of bulky waste.[26]

For consumers, discarding consumer durables meant putting the items on the street the evening before collection day – often this was offered quarterly –, or calling municipal services and arranging a pick-up date on which municipal waste workers would come and remove the items (fig. 1). Municipal waste services, on the other hand, were ill prepared for this waste fraction and its diverse material contents. Initially, several towns and cities worked in collaboration with local scrap traders which also might salvage recyclable items from local dumps. When Bochum disposed of the items of its initial collections in local dumps,[27] citizens gathered there and took reusables back home. To avoid this circulation of abandoned goods for reasons of hygiene, the bulky waste was eventually burnt on site. Darmstadt handed the responsibility entirely to scrap dealers to organise a yearly bulky waste collection.[28] Waste salvage still constituted a business around 1960, and the scrap trade (*Altwarenhandel*) involved buying various used materials to process them for reuse in production. But such reuse channels would soon disappear as a result of declining scrap prices and increasing combi-

---

24 See Lösch, Klaus: Probleme des Abfallaufkommens und der Abfallbeseitigung dargestellt am Beispiel bundesdeutscher Städte, Bremen: 1984, p. 24; Argus [Arbeitsgruppe Umweltstatistik] (ed.): Umweltforschungsplan des Bundesministeriums des Inneren. Abfallwirtschaft. Forschungsbericht 10303503. Bundesweite Hausmüllanalyse 1979/80, Berlin 1981, here p. 158.

25 Müll und Abfall, Beihefte 14 (1978); Langer, Hans/Stief, Klaus: Menge und Zusammensetzung von Abfällen, Berlin: Schmidt 1978, p. 37.

26 Hungerbühler, Eberhard: Neuer Rohstoff Müll-Reycling, Ravensburg: Maier 1975, p. 39.

27 Mahlke, Sperrmüllabfuhr, p. 395–397.

28 Erbel, Sperrmüllabfall, p. 5.

nations of material components – dismantling and recycling a cast iron stove was safe and simple when compared to a 1970s television set. Most bulky waste ended up in municipal landfills, where it was shredded, bulldozed and buried along with other waste. By 1975, municipal waste consisted of roughly 65% household waste, 9% bulky waste and 26% trade waste. By now, nearly 40 incineration plants were in operation, and these likewise absorbed a share of around 9.5% of bulky waste.[29]

Figure 1. Scene from Düsseldorf's bulky waste collection: waste workers load discarded sofas, prams, chairs and mattresses for disposal. Photograph by Irmgard Baum (undated). Source: Stadtarchiv Düsseldorf, 025_422_010.

---

29 Gerhards, Kurt-Hermann: Menge und Art der Kunststoffe im Müll und ihre Bedeutung für die Schadstoffemission aus Müllverbrennungsanlagen, PhD dissertation, University of Stuttgart 1975, p. 39.

As a highly visible sign of changing usage and disposal patterns, bulky waste became a contested icon of mass consumerism and throwaway practices. For public discourse, cultural critics, concerned citizens and waste professionals alike, it indicated West Germany's rapid transition to a throwaway society. Vance Packard's *Waste Makers* – a revealing perspective on the ephemerality of American consumer culture, translated as "Die große Verschwendung" in 1961 – was widely read,[30] and in the public debate, corporate America's promotion of convenient disposability was almost univocally rejected. The author and journalist Theo Löbsack interpreted bulky waste as the "capitulation of large swathes of the population" unable to resist the lure of advertising and the latest fashions. In his eyes, discarded "radios that still played, irons that still could be ironed with" represented a consumer culture that threw away consumer durables "by sudden aversion or because of the irresistible wish to own the 'latest model'".[31] Aware of the fact that the items consumed would eventually end in municipal waste heaps, waste experts closely observed changing consumption patterns. They noted the decline of the waste salvage trade which had once scrapped old washing machines or stoves as much as the role of fashion that on average saw "one generation … consume at least two complete sets of furniture".[32] In 1970, a waste practitioner described bulky waste as "a true reflection of the development of civilisation and rising standard of living of populations …, with the first signs of the transition from a 'consumer society' to a 'throwaway society' becoming apparent".[33]

Over the course of the 1970s, bulky waste was still seen as a sign of throwaway consumption patterns, but what was more prominent in public discourse was its rediscovery as a reservoir for reusables. Salvaging piles of bulky waste for vintage finds became such a trend that some municipalities stopped announcing collection dates in the daily press. "If you search through the mountains of old cardboard boxes, you will find wall clocks, grandfather clocks and pocket watches, dressers, tables and chairs from the last 50 years of German home decor and sometimes also from the Biedermeier era", reported the news magazine

---

30 Packard, Vance: The Waste Makers, New York: McKay 1960; Packard, Vance: Die große Verschwendung, Düsseldorf/Vienna: Econ 1966 (1st edition: 1961).

31 Löbsack, Theo: "Müll-Lawine", in: Universitas 12 (1971), p. 1285–1294, here p. 1287.

32 Wienbeck, U.: "Die Entsorgung als Teilaspekt der Infrastruktur in Verdichtungsgebieten", in: Städtehygiene 6 (1970), p. 138–142, here p. 138.

33 Jäger, Bernhard: "Menge und Zusammensetzung von Siedlungsabfällen", in: Städtetag 4 (1970), p. 205–209, here p. 208.

Der Spiegel in 1972.[34] Searching through bulky waste became a hobby for refurbishing enthusiasts and was daily labour for those who provided goods for flea markets and second-hand stores when these became fashionable in the late 1970s.[35] Michael Thompson's now classic "Rubbish Theory" originates in this specific historical setting; it describes how what is worthless for one user may be highly valuable for another.[36]

In the 1980s, public discourse finally shifted to the environmental problems of bulky waste. Environmentally aware citizens rescued items of bulky waste from the dump as a way of saving resources and protesting against over-consumption. A 1990 guidebook on reusing bulky waste depicted such contemporary rag pickers as heroes, "the true nature conservationists, recycling specialists and ultimately perhaps decisive for saving the blue planet we call earth".[37] The book also gave tips on repairing old equipment and on channelling valuable items such as copper wire and kitchen sinks back into economic circulation. Moreover, by the late 1980s, the potentially hazardous effect of dumping electronic waste became evident: along with televisions, electronics and refrigerators, toxic chemicals such as PCBs and CFCs were being channelled into landfills and the air.[38]

"MENDING" CONSUMER ELECTRONICS: CONDITIONS AND CHANNELS FOR REUSE, RESALE AND (SELF-)REPAIR

For clothing, the shift from mending to ending has been frequently described – and with it, the changing conditions of repair and reuse.[39] Mending helped work-

---

34 "SPERRMÜLL. Fuzzis und Schrotties", in: Der Spiegel, 18 Sep. 1972, p. 175.
35 "Schrank ausgemistet. Auf der Mode-Szene blüht ein neuer Zweig. 'Second-hand'-Handel mit teuren Modellen", in: Der Spiegel, 24 Jul. 1978, p. 141.
36 Thompson, Michael: Rubbish Theory. The Creation and Destruction of Value, Oxford: University Press 1979; for the cult of refurbishing old cars see Lucsko, David N.: "'Proof of Life' – Restoration and Old-Car Patina" (this volume).
37 Golluch, Norbert/Klöstzer, Eckhard: Das Sperrmüll-Buch, Reinbek: Rowohlt 1990, quote p. 12.
38 Organisation for Economic Co-operation and Development (ed.): Product Durability and Product Life Extension. Their Contribution to Solid Waste Management, Paris: OECD 1982, p. 65.
39 Sudrow, Reparieren; Lockren, Patricia: "Strategien und Techniken textilen Reparierens. Eine Exploration anhand englischer Frauenkleidung des ausgehenden 19.

ing families to wear clothes at length, hand them down or reuse them, while bourgeois households valued needlework as a female skill and a sign of provident housekeeping. Tailors also offered mending services, but their number fell from 150,000 in the early 20th century to 12,000 in the late 1970s in West Germany.[40] The shift was gradual, and in the early 1960s, housekeeping books and even the popular women's fashion journal *Brigitte* featured tips on how to mend or reuse worn-out clothes.[41] Marketed as a disposable garment, nylon tights are an extraordinary example, with *Laufmaschendienste* (literally "ladder services") offering professional repairs to runs in tights until around 1970, when high labour costs and low-priced nylon tights led to a decline in the service.[42] Only some rare users subsequently resorted to workarounds such as nail polish to stop runs. But the lengthy historical (predominantly female) tradition of needlework has not vanished and can be seen in the sewing boxes still found in many households (fig. 2).

When it comes to electrical appliances, statistics and historical studies document their rapid dissemination but pay less attention to questions of repair, reuse and disposal. A rare survey from around 1970 reported the following figures on disposal, at a time when nearly three quarters of households owned a washing machine and refrigerators were in use in 84% of households:[43] a quarter of used refrigerators were given away to relatives or other users, 8 to 22% were used as a

---

und beginnenden 20. Jahrhunderts", in: Technikgeschichte 79/3 (2012), p. 291–298; Derwanz, Heike: Zwischen Kunst, Low-Budget und Nachhaltigkeit. Kleidungsreparatur in Zeiten von Fast Fashion, in: Krebs, Stefan/Schabacher, Gabriele/Weber, Heike (eds): Kulturen des Reparierens, Dinge – Wissen – Praktiken, Bielefeld: transcript 2018, p. 197–224.

40 See Lenger, Friedrich: Sozialgeschichte der deutschen Handwerker seit 1800, Frankfurt a. M.: Suhrkamp 1988, p. 178; König: Wegwerfgesellschaft, p. 71.

41 See e. g. Richter, Else: Das große Haushaltsbuch, Gütersloh: Bertelsmann 1964, p. 552; For "Brigittes Modebriefkasten" see e. g. Brigitte, 12 Jun. 1959, p. 33 or ibid., 5 Feb. 1959, p. 14.

42 Weber, Heike: "Made to Break? – Lebensdauer, Reparierbarkeit und Obsoleszenz in der Geschichte des Massenkonsums von Technik", in: Krebs/Schabacher/Weber, Kulturen des Reparierens, p. 49–83.

43 Study quoted in: Fleischer, Arnulf: Langlebige Gebrauchsgüter im privaten Haushalt. Ein Beitrag zu Bedarfsentwicklungen privater Haushalte unter besonderer Berücksichtigung des Ersatzbedarfs, Frankfurt a. M.: Peter Lang 1983, p. 257–258. For figures on distribution, see Wölfel, Sylvia: Weiße Ware zwischen Ökologie und Ökonomie. Umweltfreundliche Produktentwicklung für den Haushalt in der Bundesrepublik Deutschland und der DDR, Munich: oekom 2016, p. 84.

means of payment to buy a new one, and some were kept as a second or spare refrigerator. Around 30% were discarded, and two thirds of those still worked. And while in 1966, only one in every five old washing machines that were replaced were discarded, this figure rose to 70% a few years later – most likely because of a significant innovation of that time, as fully automatic models made former semi-automatic washers obsolescent.

Figure 2. The inherited sewing box of Christa Weber, the author's mother.

These figures indicate the persistence of handing down and resale; such private or professional channels for disposal and reuse co-existed alongside the discarding of bulky waste. Most specialised shops not only sold new wares but also offered repair services, and many also sold or otherwise reused old equipment – a fact which was never reflected in trade statistics. The tendency of retailers to sell used equipment for payment gradually dwindled at different times for different product categories. Around 1985, formal second-hand offers represented 15% of cameras sold and nearly 6% of television sets, while the sale of used domestic appliances was rare.[44] Second-hand washing machines, for instance, accounted

---

44 Gebhardt, Peter: Der Markt gebrauchter Güter. Theoretische Fundierung und empirische Analyse, Hamburg: Kovač 1986, p. 70, 77 and 85.

for only around 3% of the yearly sales in that product category.[45] Since then, complex safety and liability regulations for second-hand electrical equipment have made formal resale and refurbishment more complicated and, except from rare models known for their durability, exports dominate the trade.[46]

From a user viewpoint, refurbished or repaired second-hand models at reduced prices enabled consumers with lower incomes or young people setting up home to acquire otherwise unaffordable equipment. From the 1970s onwards, rummaging through second-hand stores or bulky waste heaps also became a fashionable hobby. Moreover, next to hoarding old things in cellars or storerooms, cabinets or drawers, private resales or handing down equipment remained prevalent practices, as indicated by the classified ad journals that appeared in the local press in large cities in the 1970s or current online portals such as eBay. There was a thriving second-hand trade in West Berlin, which represented a unique case because of its isolated geographical position. The weekly local newspaper *Zweite Hand* ("Second Hand") was full of private resale offers, certain districts were known for their bric-a-brac shops, and the trade telephone directory from 1989 listed 262 antique shops, 85 second-hand dealers and over 40 clearing-out businesses.[47]

The mass production and consumption of electrical appliances gave rise to approximate rules of thumb for average lifespans (which were rarely made explicit) and a demand for regular maintenance and repair. While an analysis of the content of bulky waste shows that broken appliances were not necessarily "mended" any more, the repair sector for domestic electrical equipment flourished on the basis of the growing quantities sold. In the 1970s, experts noted that refrigerators regularly broke down in the first years after purchase but that they lasted for around 10 to 12 years.[48] Black-and-white television sets were said to

---

45 Ohlwein, Martin: Märkte für gebrauchte Güter, Wiesbaden: Deutscher Universitätsverlag 1999, p. 39.

46 E.g. Miele washing machines or Vorwerk vacuum cleaners, see Broehl-Kerner, Horst et al. im Auftrag des Umweltbundesamtes: Second Life. Wiederverwendung gebrauchter Elektro- und Elektronikgeräte, Berlin: Umweltbundesamt 2012.

47 Klocke, Andreas/Spellerberg, Annette: Aus zweiter Hand. Eine sozialwissenschaftliche Untersuchung über den Second-Hand-Markt in Berlin/West, Berlin: Berlin-Verlag Spitz 1990; Janßen, G./Burkard, T.: Recherche über Projekte und Initiativen, die sich mit der Weiterverwendung von Sperrmüll befassen, Manuskript, Berlin 1989, p. 2.

48 Upmalis: "Lebensdauer von Haushaltsmaschinen und -geräten", in: HLH. Lüftung, Klima, Heizung, Sanitär. Zeitschrift des Vereins Deutscher Ingenieure 33/12 (1972), p. 389f.

need on average one repair per year and to have a lifespan of 6 to 12 years; for portable radios (and cassette recorders) these figures were respectively 0.8 or 1.3 repairs a year and a lifespan of 4 to 10 or 3 to 7 years.[49] To avoid transporting bulky items, technicians often repaired them in the home. In the case of Siemens, the 55 million electrical appliances installed in West German households in the late 1970s resulted in 1.8 million annual repairs, 90% of which took place in the home. Repair services were provided by Siemens customer services, contracted repair shops or specialist shops.[50] Among consumers surveyed at that time, only eleven percent had never required a repair service for their radio, television or record player.[51]

At the same time, some consumer studies remarked a lack of willingness among consumers to have things repaired since searching for repair facilities could be demanding and consumers often overestimated repair costs.[52] So these general assertions have to be differentiated for different product categories. For upscale consumer electronics – a term which was growing in popularity – surveys suggest that consumers decided where to buy depending on the availability of customer and repair services or customer advice.[53] Until the 1970s, customer services were still operating on a local basis and evaluated by word-of-mouth[54] – stores were either known or notorious for their repair services or warranty options. According to a survey from the late 1970s, the majority of consumers favoured specialised stores for purchase or repair; only a third of consumers bought their consumer electronics in warehouses, mail-order outlets or wholesale stores.[55] This situation would change fundamentally once consumer

---

49 Rühl/Hantsch, Strukturuntersuchung, p. 94.

50 Tietz, Bruno: Der Markt für Haushaltselektrik und -elektronik in der Bundesrepublik Deutschland von 1960 bis 1990 (Symposium 26. bis 27. Jan. 1979, eine Branchenanalyse), Hamburg/Saarbrücken: Gruner und Jahr 1979, p. 385.

51 Quoted in Rühl, Günther/Hantsch, Georg (eds.): Strukturuntersuchung in den Elektrohandwerken, vol. III: Absatzmarktentwicklung in Zahlen, Karlsruhe: ITB (Institut für Technik der Betriebsführung im Handwerk) 1979, p. 218.

52 E.g. reported in: Clemens, Brigitte/Joerges, Bernward: Ressourcenschonender Konsum. Sozialwissenschaftliche Aspekte häuslicher Abfallproduktion und -verwendung, Berlin: Wissenschaftszentrum 1979.

53 Rühl/Hantsch, Strukturuntersuchung, p. 218–219 and 209; Tietz, Markt, p. 1257.

54 In the late 1970s, however, the German Crafts Association (Zentralverband des Deutschen Handwerks) suggested establishing regional centres that would serve as intermediaries between repair services and consumers. See Tietz, Markt, p. 861.

55 Tietz, Markt, p. 380.

electronic chain stores such as MediaMarkt and Saturn offered attractive prices at the expense of customer service and repair from the 1980s onwards.

können Sie dieses nachziehen und beide Hälften vollständig voneinander trennen.

5. Der Stufenschalter liegt nun lose im Gehäuse; er ist nur noch durch die angelöteten Kabel und die ihn umfassende Kunststoffhalterung in einer der Gehäusehälften fixiert. Wenn Sie die Halterung entfernen, wird auch der Auswerfer zugänglich, den Sie dann ebenfalls herausnehmen können.
Schauen Sie sich nun den Schalter näher an. Sind die Kontakte korrodiert, können Sie diese problemlos mit Schmirgelpapier reinigen.

6. Unter Umständen finden Sie eine gelöste Lötstelle, die neu gelötet werden muß (vgl. dazu Grundkurs »Ab- und Anlöten von elektrischen Kontakten«, S. 36). Verfolgen Sie auch den Kabelweg zum Eingang des Netzkabels ins Gerät (Lüsterklemmen) und überprüfen Sie die Anschlüsse (eventuell nachziehen).

7. Die Schalterfunktionen überprüfen Sie mit dem Universal-Multimeter. Messen Sie die einzelnen Stufen mit dem Ohmmeter durch. Zwischen Stufe 2 und 3 ist häufig eine Diode zwischengelötet. Ist diese defekt, funktioniert nur noch die Schaltstufe 3.

8. Ist der Fehler festgestellt und behoben, wird das Gerät noch gründlich gereinigt und die einzelnen Lager mit etwas Öl versehen. Zum Überprüfen und gegebenenfalls Auswechseln der Schleifkohlen verfahren Sie wie im Grundkurs »Auswechseln der Kohlen bei stark beanspruchten Motoren«, Seite 38, beschrieben.

9. Achten Sie beim Zusammenbauen des Geräts unbedingt darauf, daß keine Kabel gequetscht werden und alle beweglichen Teile, insbesondere der Schalter, an den dafür vorgesehenen Stellen liegen. Vergessen Sie auch nicht, die Zugentlastung des Gerätekabels wieder anzuziehen. Der Zusammenbau erfordert etwas Fingerspitzengefühl.
Wenn nun die beiden Gehäusehälften zusammengesetzt sind, wird durch die Dämpfungsgummis etwas Widerstand erzeugt, so daß die beiden Gehäusehälften nicht ganz dicht aufeinanderliegen. Beim Anziehen der Schrauben müssen Sie jedoch darauf achten, daß die Hälften absolut deckungsgleich sind.

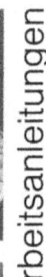
7

8

9

Figure 3. Cleaning the components of a mixer to enhance its longevity. Self-repair guides in the 1980s considered the maintenance of electrical appliances as a male task, while today's iFixit video clips also display many female hands. Source: Middel, Bernd/Müller-Steinborn, Martin: Selbst Haushaltsgeräte warten und instand setzen, Munich: Compact 1989, p. 71–73, here p. 73.

But to what extent did consumers reach for the wrench themselves? Along with do-it-yourself, self-repair became a trend in the post-war decades that eventually filled up domestic toolboxes – the male pendant to the sewing box.[56] Repair

---

56 Kreis, Reinhild: Selbermachen. Eine andere Geschichte des Konsumzeitalters, Frankfurt a. M./New York: Campus 2020; Voges, Jonathan: "Selbst ist der Mann": Do-it-

booklets, all clearly written with male readers in mind, first advised that broken electrical gear should be left to specialists for safety reasons and because special tools were needed.[57] Indeed, according to a survey in the late 1970s, only a minority of users indicated that they were able to carry out repairs themselves.[58] Although many users fixed simple problems, only technical enthusiasts used soldering irons or voltmeters. The guidebook on bulky waste mentioned above encouraged readers to salvage discarded electrical appliances for self-repair, as they often simply had broken plugs or switch buttons, or the V-belts in electrical engines may have just slipped.[59] Repair guides of the 1970s and 1980s suggest that users of that time still carried out maintenance that current readers will most likely have forgotten about (fig. 3): they instructed their envisioned male readership on how to disassemble hair dryers, electric razors or blenders so that they could take them apart and clean the parts – routines which made appliances last longer.[60]

GROWTH AND DECLINE: PROFESSIONAL REPAIR AND RESALE OF TELEVISIONS AND RADIOS

Professional repair takes place at a complex intersection where users, producers and suppliers of spare parts and toolkits, retailers, and repairers meet. Over time, repair has had to deal with rising labour costs, falling costs for new items, and changing materials and technologies. As an economic sector, repairing electronic devices has high entry barriers when compared to shoe or clothes repair, a sector in which underprivileged or untrained individuals can set up a business, often in combination with a second job. Electrical repair is regulated by professional and trade infrastructures, national safety and resale regulations and supplier conditions for spare parts. On a macro scale, national or transnational consumer policy and economic as well as technological parameters such as accelerating innova-

---

yourself und Heimwerken in der Bundesrepublik Deutschland, Göttingen: Wallstein 2017.

57 Fellensiek, Hans: Selber reparieren – aber wie?, Cologne: Buch und Zeit Verlagsgesellschaft 1964, p. 39.

58 In the study quoted by Tietz (without further reference), only around 2 to 6% claimed to repair their television, radio or cassette recorders, see Tietz, Markt, p. 380.

59 Golluch/Klöstzer, Sperrmüll-Buch, p. 12.

60 Middel, Bernd/Müller-Steinborn, Martin: Selbst Haushaltsgeräte warten und instand setzen, Munich: Compact 1989.

tion cycles, the increasing miniaturisation of products and components and the ever growing variety of product models all play a part. While this section collates television and radio repair, it also explores their heterogeneity and the parallel upswing of television repair and downswing of radio repair.[61] And while the German electrical repair sector has shrunk since the late 1980s, once Asian manufacturers and outsourcing strategies had massively transformed the market, electrical repair services have boomed in the Global South.

In West Germany, in 1967, there were roughly 6,000 active radio and television enterprises, employing a workforce of around 34,700 people; in 1976, over 7,600 enterprises provided work for 39,600 employees,[62] among them many radio and television technicians. This was a registered craft; technicians were authorised both to sell and to repair or install domestic sets and aerials, and they were valued for this combination, as shown in the previous section. Moreover, in the heydays of the radio and television repair sector, i. e. in the 1950s and the 1960s and 1970s respectively, repair-only services were also active. Around 1980, for instance, West Berlin's trade phone book listed over eight pages of "television repair services", among them a few full-page advertisements offering instant assistance.[63] In the late 1970s, an estimated 20% of consumer spending on radio, television and audio equipment was for repair and servicing and 80% for new acquisitions.[64]

As with cars,[65] the first generations of televisions were known for their error-proneness. The television's test pattern, broadcast from 1950 to 1997, was an obvious sign of the need for regular television maintenance – it helped technicians to check functions such as sharpness and colour matching. Substituting delicate tubes with transistors made televisions (and radios) more reliable. But with the transition to colour technology in the 1960s, lifetimes even declined for

61 Krebs, Stefan/Hoppenheit, Thomas: "Questioning the Decline of Repair in the Late 20th Century: the Case of Luxembourg, 1945–1990", in: Hilaire-Pérez, Liliane et al. (eds.): Technical Cultures of Repair from Prehistory to the Present Day, Turnhout: Brepols Publishers 2021 (forthcoming).
62 Tietz, Markt, p. 837.
63 See Branchen-Fernsprechbuch 1980/81, Berlin West, p. 399–407, online: https://digital.zlb.de/viewer/image/15849345_1980-81/1/ (accessed 15.01.2020).
64 According to interviewed experts which were participants at the yearly conference of the "Bundesfachgruppe" for radio and television technicians. See: Rühl/Hantsch, Strukturuntersuchung, p. 94.
65 Krebs, Stefan: "Maintaining the Mobility of Motor Cars: The Case of (West) Germany, 1918–1980" (this volume).

some years owing to unreliable novel electronics.[66] Moreover, colour television designs required three times as many components as black-and-white sets. As a consequence, the repairer's toolkit had to include more instruments, and repair knowledge and skills had to be updated.[67]

By the mid-1960s, over 60% of West German households had black-and-white televisions; colour TVs began to be distributed more widely in the 1970s.[68] In 1961, the German manufacturer Telefunken reported that 80% of televisions purchased needed repairs within a year, and by 1975, the Stiftung Warentest, the nation's leading consumer testing institution, was still warning consumers that colour TVs were most likely to develop faults within their first half year of use.[69] Suppliers worked closely with licensed repair workshops to mitigate this situation, but they only gradually introduced warranty systems, while some manufacturers like Graetz with its *Prüfgarantiekarte* made internal product testing the norm. By the 1970s, warranties for TV sets still covered only material costs and not the costs of repair. Insurance companies thus developed special policies to cover damage to television sets, but their offers were too expensive to prove popular.

Customer service and professional repair were key for the widespread uptake of televisions. Given the long-term vulnerability of the technology, suppliers supported and expanded repair services which had originated in the 1950s with the radio repair and engineering profession. Suppliers, electrical guilds and professional associations offered training courses to build up a knowledge base and recruit service technicians.[70] By contrast, the contemporary radio sector was a mature market that was now dominated by replacement purchases. In this field,

---

66 Teupe, Sebastian: Die Schaffung eines Marktes. Preispolitik, Wettbewerb und Fernsehgerätehandel in der BRD und den USA 1945–1985, Berlin: De Gruyter 2016, p. 105.

67 Knobloch, Winfried: Prüfen, Messen, Abgleichen. Service an Farbfernsehempfänger. PAL – SECAM, Berlin: Verlag für Radio-Foto-Kinotechnik GmbH 1970, p. 5 and 27; see also Knobloch, Winfried: Prüfen, Messen, Abgleichen. Fernsehempfänger-Service. Berlin: Verlag für Radio-Foto- Kinotechnik 1962, p. 5 and 20.

68 Rühl/Hantsch, Stukturuntersuchung, p. 101.

69 Teupe, Schaffung, p. 123; see also: "'test'-Report: Kundendienst bei Farbfernsehgeräten", in: Radio-Fernseh-Händler 11 (1975), p. 16.

70 Fellbaum, Günther: Fernseh-Service-Handbuch. Ein Kompendium für die Berufs- und Nachwuchs-Förderung des Fachhandels und Handwerks, 2nd edition, Munich: Franzis 1962, p. 209; Hewel, Horst: Fernseh-Service-Lehrgang, in: Funk-Technik 8, 2 (1953), p. 17–18; Anon.: "Neue Philips-Fernschlchrgänge beginnen", in: Radio-Fernseh-Händler 10 (1964), p. 406.

repair and repairability were soon to become the exception. Already around 1960, reference books for radio repairmen were noting that radios were no longer being designed with the needs of repairmen in mind, that designs were complicating repair and that producers were forcing repairmen to use only original components.[71] Transistor pocket radios, unique at the time, materialised the trend which would eventually drive the overall consumer electronics market: producers disregarded repairability in favour of fashionable, fast changing designs, miniaturisation and cheap mass production methods.

By 1963, West German consumers could choose from over a hundred models in the pocket radio segment alone; for producers, annual model changes were the norm, and cheap Japanese offers had entered the market.[72] When testing pocket transistor radios in 1964, the consumer magazine DM – West Germany's leading magazine for evaluating consumer products – warned that several of them provided neither customer service nor spare parts, but were "built to be thrown away".[73] While they were available for 20 DM, a West German product like Telefunken's pocket transistor radio *Partner N* cost 156 DM. Pocket radios also illustrated the link between miniaturisation, mass production methods and repairability: shrinking sizes, the integration of components and housings with bonded joints instead of threads meant that individual breakage points were virtually unrepairable.[74]

The overall challenge facing repair services lay in shifting cost ratios which meant that the repair of equipment of average or inferior quality no longer made economic sense. Labour costs increased while prices of new appliances fell. Moreover, these new appliances soon lost in economic value. The valuation tables on consumer electronics which were compiled for second-hand merchants clearly illustrate this accelerated depreciation: the 1964 price listing for used radios, TVs and tape recorders, for instance, described a 10-year-old portable radio

---

71 See e.g. Knobloch, Prüfen; Renardy, Adolf: Leitfaden der Radio-Reparatur, 2nd edition, Munich: Franzis 1958, p. 281; Renardy, Adolf: Radio-Service-Handbuch. Leitfaden der Radio-Reparatur für Röhren- und Transistorgeräte, 3rd edition, Munich: Franzis 1963, p. 231.

72 Weber: Heike: Das Versprechen mobiler Freiheit. Zur Kultur- und Technikgeschichte von Kofferradio, Walkman und Handy, Bielefeld: transcript 2008, p. 110.

73 See "Taschenradios", in: DM 50 (1964) p. 41–46, here p. 43.

74 On encasing see Weber, Heike: "Blackboxing? – Zur Vermittlung von Konsumtechniken über Gehäuse- und Schnittstellendesign", in: Bartz, Christina/Kaerlein, Timo/Miggelbrink, Monique/Neubert, Christoph (eds.): Gehäuse. Mediale Einkapselungen, Paderborn: Wilhelm Fink 2017, p. 115–136.

as "barely worth the scrap value".[75] A *Partner N* transistor radio bought in 1961–62 was worth less than a sixth of its original price, and televisions that were more than seven years old were not listed at all. Even devices which were only four or five years old could be sold by second-hand dealers for only small sums of money. Two decades later, this semi-official listing for resellers included only TV, radio, tape and video recorder models from the past six years. By contrast, along with rising wages which were the basis of increasing consumption, professional repair was becoming more expensive. One hour of television repair work amounted to 9 DM in labour costs in 1962, 18 DM in 1971, 40 DM in 1980 and even 50 DM in 1984 – not including materials or call-out fee.[76] A 1972 handbook on the professional repair of electrical appliances suggested that there was no point in troubleshooting in cases where repair costs would exceed the costs of buying a new model or where replacement parts would soon no longer be available; according to this professional handbook, "the throwaway method already applies to many components today".[77]

CRITICISING THE THROWAWAY SOCIETY AND THE 1970S PLANNED OBSOLESCENCE DEBATE

This last section focuses on the 1970s consumer criticism, and more specifically on the reproach that industry willingly shortens product lifespans.[78] As will become clear, these debates placed obsolescence and throwaway habits in their wider economic and sociocultural contexts, incorporating factors such as economic competition, innovation politics, securing or endangering jobs and economic growth respectively, and consumer behaviour.[79] On the other hand, the discourse barely mentioned the environmental potential of reuse and repair; it failed to give consumers a voice of their own and to elucidate the critical issue of how producers and users co-determine the length of time for which products are

---

75 Döpke, Heinrich (ed.): Bewertungsliste für gebrauchte Rundfunk-, Fernseh- und Tonbandgeräte 1964/65, Munich: Franzis 1964, p. 3 (foreword); for the following: Taxliste 86. Bewertungsliste für gebrauchte Fernseh-, Rundfunk-, Tonbandgeräte und Videorekorder, 33rd edition, Munich: Franzis 1986.

76 Teupe, Schaffung, p. 127.

77 Eiselt, Josef: Fehlersuche in elektrischen Anlagen und Geräten, Munich: Pflaum 1972, p. 27.

78 See in more detail: Weber, Made to Break?

79 See also Schlotter, Hans-Günther: "Geplante Obsoleszenz als Gegenstand der Wirtschaftspolitik", in: Das Wirtschaftsstudium 5 (1976), p. 65–70.

used. Moreover, the frequent calls for enhanced product repairability were never answered.

Many prominent post-war writers expressed their indignation at mass consumerism and fashion. In the late 1950s, the Austrian philosopher Günther Anders, for instance, debunked serial products as incorporating intentional "frailty" or even having a planned "expiration date, at least approximately, and always as early as possible".[80] Writing in the mid-1960s, the German sociologist Hans Freyer defined the industrial system as surging flows of consumption in need of regular discharge, meaning that having things repaired amounted to a "sabotage" of the "production apparatus".[81] The broad take-up of Vance Packard's work also resulted in the popularisation of his distinction between three variants of product obsolescence, namely functional, qualitative and psychological obsolescence. By the early 1970s, in addition to the more traditional consumer criticism, the Marxist position denounced consumer fashion and obsolescence. For the philosopher Wolfgang Fritz Haug, for instance, commodities came "into the world with a kind of time fuse that will trigger their inner self-destruction in a calculated length of time".[82] Even a folklorist writing about German material culture noted that contemporary inventories were becoming increasingly short-lived and aligned with technological progress: instead of pianos, households owned jukeboxes, which in turn were already obsolete and would soon be replaced by the latest electronic device.[83] The subject of household effects discarded too early via bulky waste collections also made it into a children's book in 1973: the title described a "Mr Kringel" who rescued reusable items, much to the delight of the kids in the neighbourhood.[84]

The Western European transition to a mass consumer society was accompanied by steady intellectual and conservative criticism of mass consumption and wastefulness, often making use of the despised term "throwaway society". When

---

80 Anders, Günther: "Die Antiquiertheit der Produkte", in: id.: Die Antiquiertheit des Menschen, vol. 2: Über die Zerstörung des Lebens im Zeitalter der dritten industriellen Revolution, Munich: Beck 1987, p. 38–57, quotes: p. 38 and 42.

81 Freyer, Hans: Schwelle der Zeiten. Beiträge zur Soziologie der Kultur, Stuttgart: Deutsche-Verlags-Anstalt 1965, p. 223–252.

82 Haug, Wolfgang Fritz: Kritik der Warenästhetik. Gefolgt von Warenästhetik im High-Tech-Kapitalismus, Frankfurt a. M.: Suhrkamp 2009 (completely revised edition, first published in 1971), p. 64.

83 Freudenthal, Herbert: "Volkskundliche Streiflichter auf das Zeitgeschehen", in: Beiträge zur deutschen Volks- und Altertumskunde 10 (1966), p. 119–133.

84 See Mitgutsch, Ali: Warum macht Herr Kringel nicht mit? Ravensburg: Ravensburger-Buchverlag 1973.

the Club of Rome's *Limits to Growth* study (1972) disclosed the scarcity of global resources and the oil price crisis shattered the Western belief in unlimited economic growth, this was supplemented by environmental arguments. Only from the 1980s onwards was mass consumerism evaluated in novel terms and appreciated as a source of meaning or identity formation and a basis for national economic well-being.[85]

With regard to planned obsolescence, the 1970s public discourse saw both consumers complaining about consumer durables that did not live up to expectations[86] and technicians reporting that the service life of durables had become "shorter compared to past experience".[87] The car industry was even openly reproached for incorporating predetermined breaking points such as overly thin and corroding body sheets. These vague reproaches culminated in an academic debate following the publication of two officially commissioned studies on the issue. The "Commission for Economic and Social Change", a think tank set up in 1971 to guide the social-liberal West German government in its future economic and social policies, commissioned these studies to elucidate the matter and to settle the question of whether planned obsolescence represented a real problem.

The first study interviewed scientists, managers and engineers as well as sales clerks in the electrical, automotive and aviation industry on the subject.[88] Many of them saw planned obsolescence as an excessive phenomenon of the market economy and industrial interests, but also as a result of producers' alignment to consumer behaviour. As one manager stated, consumers did not want a product "for their lifetime". Sales clerks in particular emphasised the role of obsolescence in maintaining economic growth and thus jobs.

It was the second report, authored by the economist Burkhardt Röper, which sparked off an intense debate inside academia which extended to the public, but

---

85 Trentmann, Frank: "Unstoppable. The Resilience and Renewal of Consumption after the Boom", in: Doering-Manteuffel, Anselm/Raphael, Lutz/Schlemmer, Thomas (eds.): Vorgeschichte der Gegenwart. Dimensionen des Strukturbruchs nach dem Boom, Göttingen: Vandenhoeck & Ruprecht 2016, p. 293–307.

86 Mentioned e.g. in Schlotter, Obsoleszenz, p. 68

87 Mentioned e.g. in Upmalis, Lebensdauer, p. 389.

88 Barck, Klaus/Mickler, Otfried/Schumann, Michael: Perspektiven des technischen Wandels und soziale Interessenlage. Eine empirische Untersuchung über die Einstellung zum technischen Wandel von Spitzenmanagern, Naturwissenschaftlern und Ingenieuren aus industrieller Forschung und Entwicklung und kaufmännischen Angestellten der Industrieverwaltung, Göttingen: Schwartz 1974, in particular p. 79–89, 95–107; quotes: p. 95 and 102.

less so to the political sphere. According to that report, planned obsolescence only existed in the eyes of a few social critics; rather, current lifespans were seen to correspond to consumers' wishes, industry's need for novelty and a fading object stewardship.[89] The study had many flaws, however; it lacked distinct methodology or definitions and failed to deliver adequate empirical details on prospective or real service lives.

Röper reassured industry that, by and large, it was developing optimal lifespans according to economic needs and future technical and social changes. His enquiries among manufacturers, authorities and associations had not revealed any curtailing of lifespans to the disadvantage of consumers. Remaining in the realm of hearsay, he referred to car lifespans that had risen to eight to ten years, even if consumers drove more and cared less about car maintenance, or to ephemeral nylon tights favoured by women for their superior fit. When it came to electrical appliances, Röper noticed that producers valued "technical refinement" more than long lifespans; cheap or simple models were throwaway articles since repair was too expensive. The author underlined producers' interest in frequent model changes because of the high rate of innovation and advised them to communicate the advantages of new models more explicitly to consumers. Curiously enough, Röper also listed some measures that could prevent "undesirable" psychological obsolescence, i.e. items considered as obsolete due to changes in consumer fashion: tax incentives could promote second-hand markets and discourage nondurable products or environmentally harmful designs; research incentives could stimulate repairable designs. Consumer organisations should provide more information about product quality, and ample consumer education should guide users towards "careful handling", "beginning with cars' maintenance and ending with care for shoes".

Critics insisted that a "desirable" obsolescence – in the name of technical progress – could not be discerned on objective grounds from an unethical, "wasteful" obsolescence. Others remarked that different interests were at stake for different people. While middle-class consumers, for instance, had the economic means to substitute durables according to style and fashion tastes, low-income households depended on affordable basic equipment; planned obsolescence would hit them harder, but cheap or second-hand offers also enabled them to participate in consumption. From a Marxist perspective – which we might re-

---

89 Röper, Burkhardt: Gibt es geplanten Verschleiß? Untersuchungen zur Obsoleszenzthese, Göttingen: Otto Schwartz & Co. 1976 (Kommission für wirtschaftlichen und sozialen Wandel, Bd. 137). For the following quotes see: p. 250–251, 327 and 329–330.

late to today's degrowth position –, obsolescence and a "hypertrophy of dura-bles' production" inevitably resulted from capitalism.[90] The consumer sociolo-gist Karl-Heinz Hillmann discussed obsolescence in the context of future socie-tal and environmental challenges:[91] short product lives would result in higher material extractions and cause more waste and toxic emissions. Moreover, they would exacerbate existing societal crises: employment and creativity would be misdirected and financial resources would be channelled into production rather than to the economic growth of developing countries and more global social jus-tice.

To fight obsolescence, Hillmann insisted that consumers needed to be em-powered.[92] Institutionalising consumer advocacy and consumer education could help buyers to value quality over price. He encouraged sociology experts to ana-lyse how obsolescence relates to society's values, interests and needs and to gather data on lifespans. But just as consumers would have to change their choices, producers would have to assume more environmental and social respon-sibility. Moreover, the state should act against market concentration and imple-ment tax incentives to steer the environmental impact of consumption and prod-uct lifespans. The author called for fixed minimum quality standards and manda-tory product testing of any new products. Like other authors, he also called for a product labelling system that would indicate tested durabilities.

Calls for long-lasting and repairable designs were widespread in the 1970s, but they never became more than a sideline in late 20th-century environmental policies. The *Limits to Growth* study identified enhanced durability of consumer durables as a way of achieving enhanced resource efficiency; West Germany's ambitious national waste policy programme in 1975 spoke in favour of durable and repairable products; the journalist and futurologist Robert Jungk pushed sci-entists to make science and technology work for thrift.[93] Within the field of de-

---

90 Bodenstein, Gerhard/Leuer, Hans: "Obsoleszenz. Ein Synonym für die Konsumgüter-produktion in entfalteten Marktwirtschaften", in: Zeitschrift für Verbraucherpolitik 5, 1/2 (1981), p. 39–50, here p. 47.

91 See Hillmann, Karl-Heinz: "Das Obsoleszenzproblem in einer Zeit der Wachstums-und Umweltkrise", in: Jahrbuch der Absatz- und Verbrauchsforschung 21/1 (1975), p. 21–45; see also id.: "Geplante Obsoleszenz. Bemerkungen zu Burkhardt Röper", in: Zeitschrift für Verbraucherpolitik 1, 1 (1977), p. 48–61.

92 See also Hillmann, Karl-Heinz: "Das Problem der geplanten Obsoleszenz aus soziol-ogischer Sicht", in: Bodenstein, Gerhard/Leuer, Hans (eds.): Geplanter Verschleiss in der Marktwirtschaft, Frankfurt a. M./Zurich: Harri Deutsch, p. 107–178.

93 Jungk, Robert: "Das Ende der großen Verschwendung", in: Manager Magazin 6 (1974), p. 100.

sign, Victor Papanek and others called for ecologically sound construction, and some West German initiatives promoted longevity, such as the Werkbund's "Long-lasting Product Foundation" (Stiftung Langzeitprodukt) or the Berlin "International Design Centre".[94]

In the 1980s, these demands from various stakeholders fell silent. When the OECD commissioned a more detailed investigation into obsolescence in the context of the waste crisis, the study, published in 1982, was unable to gather information from producers on expected lifetimes, and consumers also obviously lacked interest in the issue.[95] Although the authors were unable to retrieve information about the planned obsolescence of current consumer products, they did remark an increase in unrepairable and throwaway designs and a clear avoidance on the part of producers to actually extend product lifetimes. Indeed, product engineering in the 1980s and 1990s focused on reducing energy consumption or recyclable designs and materials rather than repairability and durability.[96] Even consumer tests barely reviewed a product's repairability and durability.[97] It was not until 1993 that the Stiftung Warentest finally started to test the durability and lifespans of washing machines.[98]

---

94 See Papanek, Victor: Design for the Real World. Human Ecology and Social Change, New York: Pantheon Books 1972; Madge, Pauline: "Design, Ecology, Technology: A Historiographical Review", in: Journal of Design History 6 (1993), p. 149–166; for the West German initiatives see Hirtz, Georg/Klose, Odo: "'Modern' als Handelsware. Plädoyer für Langzeitprodukte. Qualität statt Quantität", in: Werk und Zeit – Forum 1 (1976), p. 1–3; Anon.: "Umweltfreundliche und rohstoffbewußte Produktgestaltung", in: Städtetag 1979, p. 210–212.
95 Organisation for Economic Co-operation and Development, Product Durability, p. 24.
96 Wölfel, Weiße Ware.
97 When designers and the Stiftung Warentest were involved in product quality testing in the late 1970s, they agreed that designers were not responsible for a "better environment". In that series, only pocket cameras and portable radios were tested for "durability" and whether casings could be opened easily. See Stiftung Warentest/Rat für Formgebung (eds.): Warenqualität – Technik und Form geprüft und bewertet, Berlin/Darmstadt: Stiftung Warentest 1977, p. 16.
98 See Prakash, Siddharth et al.: Einfluss der Nutzungsdauer von Produkten auf ihre Umweltwirkung. Schaffung einer Informationsgrundlage und Entwicklung von Strategien gegen 'Obsoleszenz'. Zwischenbericht: Analyse der Entwicklung der Lebens-, Nutzungs- und Verweildauer von ausgewählten Produktgruppen, Dessau-Roßlau: Umweltbundesamt 2015, p. 71.

CONCLUDING REFLECTIONS: LESSONS FROM THE PAST FOR AN ECOLOGICAL FUTURE FOR REPAIR AND REUSE

In the 1960s and 1970s, practices of mending and ending changed substantially, along with consumer offers, product designs, and values and norms associated with consumption and wastage. In affluent societies, mass production and mass consumption led to ever increasing domestic acquisitions and a rise in personal belongings. The total number of goods in circulation increased hugely. This spurred the second-hand trade, kept repair services alive and enabled lower-income households to participate in mass consumption. When it came to disposing of these objects, municipal bulky waste services eventually became widespread, despite constant criticism of the "throwaway society". Introduced from around 1960 onwards, bulky waste services contributed to high volumes being wasted, and they soon also served as a means of discarding functional yet supposedly "obsolete" objects. Currently, around a third of large domestic appliances that are replaced by a new model in Germany are still operational.[99] The disposal of growing quantities of increasingly heterogeneous objects and materials had an environmental impact which became clear in the 1980s. It was only in the early 21st century that electrical appliances began to be increasingly channelled back to producers, and a current EU framework (2020) calls for the recycling of bulky waste.

Repair, reuse and resale – temporal interventions that prolong the usage life of objects – have not disappeared in mass consumer societies, but their meaning and their share in relation to the overall volume of domestic products have changed. Strategies for the storage, repair, reuse or resale of a household's effects, previously driven by economic reasoning, have gone from being a necessity and a norm to a deliberate decision to preserve or hand down selected items. Buying new equipment eventually became cheaper than having the old equipment repaired; only some rare personal belongings are kept for a lifetime or beyond, while private, informal resales are thriving in niche areas. For most consumers (except those without the means to choose), the choice between ending and mending has come to depend on meanings and mentalities rather than on economic considerations: the symbolic meaning of the artefact or the owner's wish to make an environmental statement. And while sewing boxes and toolboxes are still passed down from one generation to the next, the objects to be repaired with the tools they house are hardly passed on any more. Self-repair has

---

99 Prakash et al.: Einfluss der Nutzungsdauer.

not vanished, but at the same time, there is no direct link between well-equipped sewing boxes and toolboxes and self-repair routines.

In the case of electrical appliances, the progressive miniaturisation of components, increasingly short innovation cycles and the proliferation of models have challenged not only self-repair but also professional repair, and with it, re-use and resale. Moreover, the case of television and radio repair also shows that the popular assertion of a universal decline in the repair business should instead be considered on a case-by-case basis: individual product categories had specific phases of growth and decline in repair, as also seen in the current business of mobile phone repair and resale.[100]

Product obsolescence has been criticised since the early days of mass consumption. But at the same time, producers and users alike have defined and co-shaped consumer durables as "finite" artefacts, even if their views have obviously differed time and again in respect to what exactly might constitute an optimal period of use. The 1970s debate on planned obsolescence reproached producers for short lifespans or intentional breaking points. It touched upon the larger contexts of product obsolescence – the paradigms of producing, using, repairing and wasting – and resulted in consumer policy and consumer research being strengthened. But it did not clarify the questions of how producers manage product lifespans and what product durability means for users and the environment. Calls to regulate product durability or to require formal product testing or labels for repairability and product lifespans were not answered. As a consequence, we still lack any reliable knowledge on the shaping of lifespans – and such knowledge is needed to make a convincing case for enhanced repairability and durability.[101] We know virtually nothing about how prospective lifespans are projected by industrial research, material testing, construction, or marketing or by consumer associations' product testing, nor about average usage times in households or repair and second-hand markets.

Past environmental policies and civic environmental activism have concentrated on household waste recycling and energy consumption – standard labels indicating the power consumption of electrical appliances, for instance, were in-

100 Nova, Nicolas/Bloch, Anaïs: Dr. Smartphone: An Ethnography of Mobile Phone Repair Shops, Lausanne: IDPURE 2020.

101 A few recent studies have tackled the issue, see Brönneke, Tobias/Wechsler, Andrea (eds.): Obsoleszenz interdisziplinär. Vorzeitiger Verschleiß aus Sicht von Wissenschaft und Praxis, Baden-Baden: Nomos 2015; Poppe, Erik/Jörg Longmuß (eds.): Geplante Obsoleszenz. Hinter den Kulissen der Produktentwicklung, Bielefeld: transcript 2019.

troduced in the late 1970s –,[102] but they have failed to explore the potential of repair, reuse and resale for "greening" mass consumption. As a result, the recent discussions on short-lived products and the environmental potential of repair and product durability can only enumerate positions, problems, arguments and potential solutions, most of which also figured prominently in the 1970s debate. While previous debates centred around cars and household appliances – consumer technologies which had recently been widely adopted –, the focus has now moved to smartphones and digital appliances. In the past, local bulky waste piles served as a contested icon; today, these have been replaced by globalised media images of toxic e-waste piles and their scrapping in the Global South.

"Don't end it, mend it!" proclaims the recent repair movement.[103] In 2021, the European Parliament voted in favour of establishing the "right to repair" and was considering a "product pass" to clarify the lifespan and repairability of products.[104] Looking to the past shows us that the right to repair is but one element in an intricate web which determines the length of time for which a product is kept in use. Choices on mending and ending are closely entwined with the economies, systems and practices of producing, using, reusing and reselling as well as discarding. Sociocultural values and norms regarding novelty, obsolescence and what objects are worth being preserved also play their part. "Repair" is not simply about repairing technology, but about "repairing" and reassembling society and consumer culture at large.

102 Weber, Heike: "Europe's Consumer-Recyclers: The Hope to 'Green' Mass Consumption by Recycling", in: Wöbse, Anna-Katharina/Kupper, Patrick (eds.): Protecting the Environment (forthcoming); Wölfel, Weiße Ware.

103 Platform21 (2009): Repair Manifesto, http://www.platform21.nl/page/4375/en (accessed 22.12.2017).

104 Mikolajczak, Chloé: European Parliament calls for ambitious right to repair, https://repair.eu/de/news/european-parliament-calls-for-ambitious-right-to-repair (accessed 18.03.2021).

The Persistence of SS France:

Her Unmaking at the Alang Shipbreaking Yard in India

Ayushi Dhawan

> Do not call me "France" ever again. France she let me down. Do not call me "France" ever again. This is my last wish. I was a gigantic boat able to cross a thousand years. I was a giant, I was almost as strong as the ocean. I was a gigantic boat. I took away thousands of lovers. I was France. What's left of it? A mooring for cormorants. Do not call me again "France" ever again. France she let me down. Do not call me "France" ever again. That is my last wish.
>
> *Le France song lyrics by Michel Sardou, 1975*[1]

When seemingly invincible technologies break down, it is a source not only of inconvenience but also of great frustration to their respective users.[2] Trips to repair

---

1   Paroles.net: Lyrics of the song Le France by Michel Sardou, https://www.paroles.net/michel-sardou/paroles-le-france (accessed 24.03.2019).

2   Early iterations of this research were presented at the "Histories of Technology's Persistence: Repair, Reuse and Disposal" workshop at the Luxembourg Centre for Contemporary and Digital History (C²DH), and the "Life Cycle of Container Ships. Global Ethnographic Explorations into Maritime Working Lives" opening workshop at the Norwegian Maritime Museum. I would like to sincerely thank Heike Weber and Stefan Krebs for their editorial enthusiasm and engaged comments that helped me clarify my thoughts. I also wish to thank my supervisor Simone Müller and the Hazardous Travels team at the Rachel Carson Center for Environment and Society for going through earlier drafts of the paper and for their continued support, advice and inspira-

professionals follow, with a hope that objects will be fixed.[3] But what if the repair process fails to restore non-functional machinery to an operational state, or if repair costs grow out of all proportion? What happens then? It should come as no surprise that most technologies do eventually wear out or become dysfunctional for a variety of political, economic, social or cultural reasons, and are then either abandoned to the process of natural degradation or transferred to disposal sites for recycling.

Seafaring vessels are the biggest of all man-made moving objects. The average commercial lifespan of these vessels is 25 to 30 years, after which time maintaining them effectively and operating them profitably becomes uneconomical for their owners. The operational period can also be shortened by a sudden economic or financial crisis. The reasons for dismantling a ship, therefore, are many and varied. Age is one of the most frequent, and overcapacity in tonnage, changing regulations concerning shipbuilding and operations, insurance constraints are some of the other reasons why ships are sold for scrapping.

In *The Cultural Biography of Things*, Igor Kopytoff argues that "biographies of things can make salient what might otherwise may remain obscure".[4] Kopytoff proposes a biographical approach as a means of understanding the agency of objects that move through space and time, quite similarly to the biographies of people. "In doing the biography of a thing", Kopytoff suggests, "one would ask questions similar to those one asks about people".[5] He asks: Where does the thing come from and who made it? What has been its career so far, and what do people consider to be an ideal career for such things? What are the recognised "ages" or periods in the thing's "life" and what are their cultural markers? How does the thing's use change with its age, and what happens to it when it reaches the end of its usefulness?[6] Taking Kopytoff's concept of birth-life-death as a

---

tion, my partner Pulkit Kapoor who kindly assisted me with the field work in Gujarat in 2018, Anne Mette Seines for generously helping me with the translation of Norwegian texts, and all the interviewees. And finally, thanks to the DFG Emmy-Noether-Programme for providing the financial support that made this research study possible.

3   Repairs are often defined as unscheduled activities that arise out of a need to eliminate faults and make broken objects useful again on an after-the-fact basis. See Krebs, Stefan/Schabacher, Gabriele/Weber, Heike (eds.): Kulturen des Reparierens: Dinge – Wissen – Praktiken, Bielefeld: transcript 2018.

4   Kopytoff, Igor: "The Cultural Biography of Things", in: Appadurai, Arjun (ed.): The Social Life of Things, Commodities in Cultural Perspective, Cambridge: Cambridge University Press 1986, p. 64–91, here p. 67.

5   Ibid.

6   Ibid.

starting point, in this chapter I examine the voyage of the *SS France*, an end-of-life ship, by tracing the events from her first use to her disposal and recycling. The story of this ship, however, reveals the limitations of this anthropomorphic metaphor, because examples from my research show the persistence of the *SS France*, or, as I describe it, the many lives of the *SS France*. She was reborn in her second life as the *SS Norway*, in her third as the *SS Blue Lady*, and underwent a subsequent unforeseeable reincarnation after demolition, with many of her material components and steel being remobilised and reintegrated into the local economy while others turned into waste legacies at the ocean's shore.

Since September 2017, I have been tracing the journeys of end-of-life vessels that end up in Alang, Gujarat for demolition. My archival research has taken me from India to the Netherlands to Norway. The following sections are based on archival sources found at the shipbreaker's office in Alang, the archives of the Norwegian Maritime Museum and Greenpeace, and the web. They range from newspaper clippings, the *SS France*'s shipping documents, bills of sale and purchase, NGO reports, Indian court proceedings, and documentaries on the *SS France* and shipbreaking in general. This data is further complemented by direct observations made at the Alang shipbreaking yards from April to August 2018. Researchers are authorised to access shipyards if due permissions are obtained from the relevant authorities. I also draw on ten semi-structured interviews of approximately 90 minutes conducted in English and Hindi with key stakeholders, including administrative port authorities, shipbreakers, workers, union workers, second-hand market traders and academics.

In the first part of the chapter, I examine the voyage of an end-of-life ship, the *SS France*, from the Global North to the Global South for scrapping. Plying the seas, variously as the *SS France*, the *SS Norway* and the *SS Blue Lady* on behalf of a string of owners, the ship's journey came to an abrupt halt as she found herself faced with prohibitively expensive repairs owing to a fatal boiler accident on board. I argue that the history of a technical artefact, in this case a ship, should not be reduced merely to production, repair, reuse, maintenance or recycling; rather it needs to encompass all these stages, since they are deeply interwoven with questions of waste and disposal.

In the next section of the chapter, I elaborate on my hypothesis by investigating the shipbreaking site where the *SS Blue Lady* was demolished in 2006–2007. Alang is located along the Gulf of Cambay, in the Bhavnagar district of Gujarat, the north-western-most state in India. Alang and neighbouring Sosiya are two local villages that lend their name to the Alang-Sosiya shipbreaking yards (ASSBY), located along a ten-kilometre stretch of coastline. The economic and social organisation of the shipyards and the reverse logistics of taking things apart are analysed as part of a process that I refer to as the unmaking of end-of-life vessels, as they are turned by arduous human labour into a variously reusable, non-reusable and toxic stream of materials.

The unforeseeable "reincarnation" of the *SS France* and the tracing of reusable materials generated from many decommissioned vessels that are sold in the second-hand markets of Alang is the subject of the final part of the chapter. The story of the *SS France*, as the markets and shipyard reveal, is not just a story of birth and death, since it does not end after demolition; instead, her steel, toxic residues and fittings are reintegrated into the local economy, much like those of many end-of-life vessels that are sent here for scrapping.

THE SS FRANCE: THE DOOMED STORY OF THE LAST ILLUSTRIOUS SHIP OF STATE

The *SS France* was a *Compagnie Générale Transatlantique* ocean liner built at the *Chantiers de l'Atlantique* shipyard in Saint-Nazaire, France, between 1957 and 1961. She was built to replace old vessels of the 1950s such as the *SS Ile de France* and the *SS Liberte*.[7] The *SS France* was a massive steel vessel – in fact the longest passenger ship ever built, at 315 m long and 33.7 m wide – with a horsepower of 160,000 kW. She maintained her leading position for almost 44 years until the construction of the *RMS Queen Mary 2* in 2004.[8] Like every other ship embarking on its maiden voyage, she was christened by her godmother, Madame Yvonne de Gaulle, wife of French President Charles de Gaulle, at a public ceremony.[9] A bottle of champagne was smashed against the shiny hull of this $80 million liner, and she was launched into the Loire River on 11 May 1960. In his inaugural speech, President Charles de Gaulle, incredibly proud of France's naval achievements, said, "The *France* is launched. It is going to marry with the sea. May this ship accomplish its destiny to carry men toward men".[10] He applauded the shipbuilding industry, represented by a series of leaders, engi-

---

7   O'Brien, Rob: Classic Liners: SS France, http://www.classicliners.net/SSNORWAY. html (accessed 16.03.2019).

8   Ibid.

9   Clip-9888: Launching of SS France, http://www.budgetfilms.com/clip/9888/ (accessed 16.03.2019).

10   Storli, Captain Jan-Olav: SS France-Construction and Launch, http://www.captains voyage.com/ncl/ss-norway/ss-france---construction.html (accessed 10.03.2019).

neers, technicians and workers who brought the *SS France*, an icon of French maritime pride, into existence.

After outfitting and sea trials, the ship entered into service in 1962. She impressed everyone around with her speed of 30 knots, cutting long transatlantic voyages from nine to five days. For the next 12 years, the *SS France* crossed the Atlantic as a speedy ocean liner, ferrying passengers back and forth, to and from Europe and the United States. However, the advent of cheap transatlantic air travel in the 1950s–1960s, increasing fuel oil prices in 1974, the launch of even faster rival vessels and the withdrawal of all state subsidies led to the ship's early retirement (fig. 1).[11]

Figure 1. The *SS France* docked at Kowloon in Hong Kong on her final round-the-world cruise as *France*, February 1974 (Photo: https://commons.wikimedia.org/wiki/File:SS_France_Hong_Kong_74.jpg, accessed 25.09.2019).

In 1974, the *SS France* was removed from service and laid up at Le Havre port, where she remained idle for the next five years. Her mothballing was met with anger and dismay by much of the French population. Michel Sardou's protest song *Le France*, quoted in the chapter's epigraph, personified the once proud ocean liner that had been betrayed by politicians, taken out of service and left

---

11 O'Brien, *Classic Liners*.

docked at the port. The controversial song topped the pop charts, with more than 500,000 copies being sold in two weeks.[12]

HER SECOND LIFE AS THE SS NORWAY

In 1979, the *SS France* was sold to Knut Kloster, owner of Norwegian Cruise Line (NCL), one of Norway's oldest shipping firms, for $18 million. NCL spent $80 million on refitting operations in Bremerhaven, Germany, and renamed her the *SS Norway* (fig. 2).

Figure 2. The *SS Norway* arriving in Southampton on her maiden voyage after her conversion to a cruise ship, May 1980 (Photo: https://commons.wikimedia.org/wiki/File:SS_Norway_on_%22maiden%22_voyage.jpg, CC BY-SA 4.0, unaltered, accessed 25.09.2019).

The ocean liner was successfully converted into one of the largest Caribbean cruise ships. An invitation letter from Eric Bye, Norway's dearest conveyors of maritime culture, demonstrates the pride Kloster had for the newly christened *SS Norway* that was on a voyage to Oslo. He said:

---

12 Lichfield, John: "The Pride of a Nation: Luxury Liner that's set to sail once more", in: The Independent, 26 Feb. 2011, p. 1.

"This is hardly an understatement when talking about the world's largest passenger ship, the former S/S France, which has virtually become a brand-new vessel, and will be offering cruises in the Caribbean. On May 3 [1980], the S/S NORWAY arrives in Oslo on a short stop. Klosters Rederi brings the ship to the country's capital solely to present it to the population of Norway before it leaves our waters for good. Because it is not likely that the S/S NORWAY will ever return to Norway, so the event in Oslo in May will thus be both a 'welcome home' and a goodbye."[13]

She entered service on seven-night cruises from Miami to the Caribbean in the 1980s and became a celebrated luxury Caribbean cruise ship. On 25 May 2003, the eventful journey of the SS Norway came to an abrupt halt as a boiler rupture in the aft boiler room (boiler no. 23) crippled her.[14] The accident occurred while the ship was docked in Miami after having just completed a week-long Caribbean cruise. There were 911 crew members and 2,135 passengers on board when the boiler explosion happened. It killed eight crew members and injured seventeen others, and damaged not just the aft boiler room but also the bulkheads, doors and doorframes three decks above on the ship.[15]

The SS Norway, NCL's only steam-powered ship at that time, was equipped with four central boilers. The boiler system of the 43-year-old ship was originally built in France and later modified in Germany in 1999.[16] The US National Transportation Safety Board (NTSB) conducted a detailed investigation of the boiler's operational and maintenance history after the accident. The report highlighted that a weld on the seam of a high-pressure drum had ruptured, releasing

---

13 Bye, Eric: Copy of Invitation Letter "VI Gar Ombord" I S/S Norway, 15 Jan. 2019, Skipsopplysningsarkiv "Norway" collection, 1980, Archive of the Norwegian Maritime Museum.

14 Sutton, Jane: "Norwegian Cruise to plead guilty in deadly blast", in: Reuters, 3 May 2008, p. 1.

15 Snyder, John: "Eight crew killed by steam-boiler explosion aboard Norway", in: Professional Mariner, 28 Feb. 2007, p. 1.

16 The SS Norway had a chequered past: in the 1980s, the ship's electrical system had failed, as a result of which she had drifted for a day. Subsequently, in 1981, a boiler room failure idled the ship yet again for a day. This was followed by a fire in the boiler room in the same year which led to the cancellation of two more cruises. In 2001, the SS Norway failed a coast guard inspection because more than a hundred problems were found, especially with regard to unfinished maintenance and repairs on the vessel. See Anon.: "SS Norway has chequered history", in: United Press International, 26 May 2003, p. 2.

20 tonnes of scalding water that turned into steam and swept through the engine room and adjacent crew quarters on the vessel.[17]

Marine boilers are one of the most essential pieces of equipment on ships. Their main job is to generate high-pressure steam to run the ship's machinery. Like any other apparatus, boilers have an estimated design life, and the degree of material fatigue they sustain is affected by how they are operated and how often and how well they are inspected, maintained and repaired. The ship's original boiler manual stated that cleaning should have been carried out every 3,000 hours.[18] The boilers were periodically inspected visually and by non-destructive tests (such as dye penetration, magnetic particle inspection or ultra-sonics). The visual inspections were challenging to execute, however, as a surveyor could not examine the drum by just putting his head through the access opening and shining a flashlight inside. As one engineer explained, "[y]ou have to go in … and it's a very, very cramped space".[19] The boilers were found to have pitting and oxygen corrosion on multiple occasions. Throughout the years of the ship's operation, any cracks found were welded but with questionable expertise. Frequent use of incorrect boiler start-up and shutdown procedures and a failure to maintain correct water chemistry in the boiler also added to significant stresses on the machinery. In 1997, a port engineer cautioned the NCL Executive in writing: "The boilers on the SS Norway have reached a state where a decision must be made".[20] The options available to resolve the problem included replacing the old boilers, completely retubing the old boilers with new economisers, and installing new automation. The company's document analysis further revealed that no nondestructive testing or internal visual inspections were carried out after 1990 for the headers of boilers 21 and 23 and after 1996 for the headers of boilers 22 and 24.

The post-accident investigations thus determined that the most probable cause of the SS Norway's boiler rupture was "the deficient boiler operation, maintenance, and inspection practices of NCL, which allowed material deterioration and fatigue cracking to weaken the boiler. Inadequate boiler surveys by

---

17 National Transportation Safety Board Washington D. C. 20594: Marine Accident Brief, https://www.ntsb.gov/investigations/AccidentReports/Reports/MAB0703.pdf (accessed 10.03.2019).

18 Ibid.

19 Ibid.

20 Ibid.

Bureau Veritas [a French company that inspected *Norway*] contributed to the cause of the accident".[21]

The crippled *SS Norway* was registered with a Bahamian flag and was once again towed from Miami to the Lloyd Werft shipyard in Bremerhaven. After arriving on 23 September 2003, she awaited the decision as to whether she would be repaired and returned to service or taken out of service altogether. NCL decided that the ruptured boiler should not be replaced, as it was difficult to determine whether the other three boilers were safe to operate any further. The repair personnel believed that only a total replacement of the boiler machinery would make her safe to return to service once again.[22] The high repair costs were deemed uneconomical, and in March 2004, NCL announced that the *SS Norway* would not return to the North American fleet. The ownership of the vessel was transferred to Star Cruises, the parent company of NCL, for potential scrapping.[23]

A RELENTLESS SEARCH FOR A MARITIME GRAVEYARD

Seafaring vessels are not just giant floatable masses of steel; they are assemblages of a wide variety of materials. Vessels built before the 1980s, in particular, were built of substances ranging from asbestos, heavy metals, polychlorinated biphenyls (PCBs), tributyltin (TBT), chlorofluorocarbons (CFCs) and radioactive materials. The *SS Norway* was therefore no exception to the prevailing norms of the time. During her construction, these materials were used in the ship's structure and each substance intended for use had a very specific role.

In 1888, asbestos was considered to be one of "nature's most marvellous productions" owing to its thermal insulation and fire-resistant properties.[24] It is most commonly found in insulation gaskets, brake linings, pipe laggings, doors and other similar items. Heavy metals such as lead, cadmium, mercury and

---

21 National, Marine Accident Brief.

22 Newman, Doug: "SS Norway: Time is Up", in: Cruise Critic, 18 Oct. 2007, p. 1.

23 NGO Platform on Shipbreaking: Star Cruises Ltd and Norwegian Cruise Lines: Deceiving Germany and Violating International Law in the Export of the SS Norway to India, http://archive.ban.org/library/Star_Cruises_Deception_Report_Final.pdf (accessed 13.03.2019).

24 Litvintseva, Sasha: "Asbestos: Inside and Outside, Toxic and Haptic", in: Environmental Humanities 11, 1 (2019), p. 152–173.

chromium are used in batteries, level switches, gyrocompasses, galvanized materials, etc.[25] PCBs are a family of chemicals that were used on board because of their great electrical insulation and fire-resistant properties. They can most commonly be detected in oils, plastics, paints and other adhesives.[26] TBT is an umbrella term for a class of organotin compounds that were used in anti-fouling paints. As biocides, they prevented the growth of algae, barnacles and other marine organisms on ships' hulls.[27] CFCs are compounds that were used as coolants in on-board air conditioning systems and refrigerators. Radioactive materials such as Americium 241 were especially used in smoke detectors installed on ships.[28] To summarise, a combination of these materials therefore protected the ship from burning and rusting and prevented an overgrown hull from micro-organisms underwater.

But owing to their harmful occupational effects on shipyard workers and the adverse impact on the environment, these substances were progressively banned from being used in the shipbuilding industry in the Global North. Some of the materials such as asbestos, heavy metals, PCBs and TBT were identified as human carcinogens. Christopher Sellers and Melling Joseph, elaborating on the specific case of asbestos, note that as strict health regulations were introduced in developed countries, the use of asbestos decreased significantly, falling by half from its peak in the 1990s. But its use continued to increase as a result of exports to developing countries like India.[29] These embedded materials, progressively recognised as hazardous over time as health and environmental regulations became more stringent, were deemed very expensive for proper decontamination in seafaring vessels by their owners.

In 2004, an interested European third party, Pierre & Vacances, explored the possibility of purchasing the SS Norway from Star Cruises. It commissioned a feasibility study to determine the amount of asbestos the ship contained and the cost of its proper decontamination. Their study estimated that over €17 million would be needed to decontaminate some of the asbestos, covering only the partition walls, insulation and briquetting.[30]

25 Poel, Marc van de: Hazardous Materials on board, http://vandepoel.nl/wp-content/up loads/2017/06/MAR1045_Boek-asbestos-digitaal_V1.pdf (accessed 12.03.2019).

26 Ibid.

27 Ibid.

28 Ibid.

29 Sellers, Christopher/Melling, Joseph (eds.): Dangerous Trade: Histories of Industrial Hazard across a Globalizing World, Philadelphia: Temple University Press 2012.

30 NGO, Star Cruises.

Meanwhile, the somewhat abandoned *SS Norway* was docked in Bremerhaven for almost two years. On 23 May 2005, the ship was allowed to leave Germany based on Star Cruises' false declaration that the vessel was headed to Singapore and was going to be reused as a floating hotel. In reality, however, it was headed to Asia for scrapping.[31]

Under the Basel Convention on the Control of Transboundary Movements of Hazardous Wastes and their Disposal (1992), the export of ships from OECD to non-OECD countries is considered as illegal traffic and prohibited by law. The Convention does not refer directly to the vessels or rigs; rather it refers to the hazardous waste contained in them. It defines "wastes" as substances or objects which are disposed of, intended to be disposed of or required to be disposed of by the provisions of national law. In the absence of specific provisions or guidance concerning the special nature of the transboundary movement of ships for the purposes of recycling or disposal provided by the Basel Convention or the parties to it, the general waste definition must be applied.[32] Ships therefore become waste when their owners have the intention to scrap or recycle them.

The 2006 European Waste Shipment Regulations extend the obligations of the Basel Convention to all waste, whether hazardous or not. The Regulation is directly applicable in all European Union states and bans the export of waste from the European Union for recovery to any location outside the OECD.[33] Since the *SS Norway* was a ship of European origin, she should not have been sold to any country outside the European Union for scrapping. Her departure from Germany was therefore a violation of international regulations enforced to control the movement of hazardous waste.

Upon reaching Asia, the *SS Norway* did not become a floating hotel, as declared and expected. Instead she ended up in Port Klang, Malaysia, the home port of Star Cruises, on 14 October 2005. During her one-year stay, the *SS Norway* was renamed the *SS Blue Lady* and prepared for scrapping. Interested

---

31 Ibid.

32 United Nations Environment Programme (UNEP), Basel Convention on the Control of Transboundary Movements of Hazardous Wastes and their Disposal: Protocol on Liability and Compensation for Damage Resulting from Transboundary Movements of Hazardous Wastes and their Disposal, https://www.basel.int/Portals/4/Basel%20Convention/docs/text/BaselConventionText-e.pdf (accessed 12.03.2019).

33 Ship Recycling, FAQs on ship and rig recycling, https://safety4sea.com/21182626-2/ (accessed 13.03.2019).

shipbreakers from India were invited by Star Cruises to Port Klang to inspect the vessel and decide on her impending fate.[34]

Star Cruises still claimed, however, that the non-operational vessel was headed to Dubai in the United Arab Emirates for repairs. The ownership of the vessel changed yet again as Star Cruises sold the SS *Blue Lady* to a Bangladeshi Liberian-registered breaker, Bridgend Shipping Ltd, for $10 million. On 16 February 2006, the vessel attempted to berth in Bangladesh. But after intensive lobbying by Greenpeace, the Bangladeshi government refused entry to the now infamous toxic ship which contained 1,250 tonnes of asbestos among other harmful substances.[35] Bridgend Shipping Ltd then sold the ship to M/S Priya Blue Industries Pvt. Ltd, an Indian ship dismantling company, for $16 million.[36] On 30 June 2006, with the help of tugs, the vessel arrived in Alang, with 14 people on board.

RACKING BRAINS: SHOULD THE *SS BLUE LADY* BE BROKEN OR RETURNED?

Gopal Krishna, an environmental activist, initially prevented the *SS Blue Lady* from entering Indian territorial waters for demolition. He filed an application to the highest judicial body, the Supreme Court of India.[37] An inspection committee was set up, and the ship was anchored at Pipavav Port, a few nautical miles away from Alang, as the court hearing was in progress (fig. 3). On 15 August 2006, amidst all the controversies, the *SS Blue Lady* was beached on the shores of Alang,[38] with the shipowner having pleaded to the court primarily on humanitarian grounds, since the monsoon season was approaching, and food supplies were running out for the crew still on board the ship. However, the final permission for dismantling was still not granted by the court to the Gujarat Maritime

---

34 Blue Lady, SS: Bills of Sale and Purchase, 3 Jun. 2018, Priya Blue Industries Pvt. Ltd collection, 2006, Office of Priya Blue Industries Pvt. Ltd, Bhavnagar, Gujarat.

35 Goossens, Reuben: NTSB Report Discloses NCL Failures re SS Norway Boiler Maintenance, https://ssmaritime.com/norway-NTSB-report.htm (accessed 11.03.2019).

36 Blue Lady, SS: Bills of Sale and Purchase, 3 Jun. 2018, Priya Blue Industries Pvt. Ltd collection, 2006, Office of Priya Blue Industries Pvt. Ltd, Bhavnagar, Gujarat.

37 Watch, Bihar: Human Rights and Environmental Groups condemn Blue Lady Ruling, http://www.asbestosfreeindia.org/2007/09/ (accessed 12.03.2019).

38 The method used for shipbreaking in Alang (and in other South Asian countries) is known as the beaching method. Ships are run aground at high tide, leaving them stranded at low tide. It is an irreversible process and after beaching the ship cannot run on its own power.

Board (GMB), the nodal agency that regulates shipbreaking activities in Guja-rat.[39]

Still hopeful environmental activists lobbied heavily for the return of the *SS Blue Lady* to Germany. At the behest of the court, a Technical Experts Commit-tee (TEC) was set up, comprising of experts and retired naval officers. They were instructed to physically inspect the vessel so that a decision could be made in the best possible manner. The TEC felt that the breaking of the *SS Blue Lady* in Alang was feasible. They argued that asbestos containment was possible as it was mostly found in enclosed areas in the form of wall partitions, ceilings, roof-ing and galleries and was reusable, as is the case with the most end-of-life ships. With proper safety equipment available to the workers, together with appropriate waste disposal facilities, demolition should be allowed.

Figure 3. The *SS Blue Lady* docked at Pipavav Port, August 2006 (Photo: Satish Singh, Safety Officer, V1 Plot, Priya Blue Industries, Alang).

On 11 September 2007, permission to dismantle the venerable ship was granted. The judgment stated:

---

39 Krishna, Gopal: "Will the Blue Lady do a Le Clemenceau?", in: India Together, 17 Feb. 2007, p. 1.

276 | Ayushi Dhawan

"We may mention that breaking of the vessel Blue Lady will provide to this country 41000 MT [metric tonnes] of steel and it would give employment to 700 workmen. [...] [T]here will be less pressure on mining activity elsewhere. [...] It cannot be disputed that no development is possible without some adverse effect on the ecology and environment, and the projects of public utility cannot be abandoned and it is necessary to adjust the interest of the people as well as the necessity to maintain the environment. A balance has to be struck between the two interests. Where the commercial venture or enterprise would bring in results which are far more useful for the people, difficulty of a small number of people has to be bypassed. The comparative hardships have to be balanced and the convenience and benefit to a larger section of the people has to get primacy over comparatively lesser hardship. [...] 85% of asbestos is in the panels and insulation that quantity is reusable."[40]

In 2006, Alang had refused to demolish *Le Clemenceau*, an asbestos-lined 27,000 tonnes French warship, and returned it to its home country after intense lobbying pressure from Greenpeace as it contained more asbestos than French officials had claimed while selling it.[41] The arrival of the *SS Blue Lady* to Indian shipyards and the court's approval to dismantle the ship could be interpreted as an attempt to revive the stagnating business in Alang, as it was receiving fewer vessels for demolition than usual.[42] In 2005, Alang received 73 vessels, compared with 361 in 1998. Competition from other shipbreaking yards in Bangladesh, Pakistan and China was seen as one of the main reasons for bad business. In his work on the *SS Blue Lady*, Federico Demaria has highlighted "different languages of valuation" used by various stakeholders involved in the shipbreaking industry, including environmentalists, shipbreakers, villagers, Indian national authorities and international authorities.[43] He has noted how different actors used different frames of reference to argue their respective cases and how language that "expresses sustainability as monetary benefit" at a national scale was dominant in the judgment of the Supreme Court of India. The journey of the *SS Blue Lady* to the Alang shipbreaking yards is one example of a ship that has reached the end of her useful life, is deemed prohibitively expensive for repair and is

40 Supreme Court of India, Research Foundation for Science Technology and Natural Resource Policy vs. Union of India and Others: Civil Writ Petition No. 657 of 1995, https://www.sci.gov.in/jonew/judis/29517.pdf (accessed 10.03.2019).

41 Jeena, Kushal: "Despair as Clemenceau returns", in: United Press International, 20 Feb. 2006, p. 1.

42 Ibid.

43 Demaria, Federico: "Shipbreaking at Alang-Sosiya (India): An ecological distribution conflict", in: Ecological Economics 70 (2010), p. 250–260.

therefore sent for scrapping as she has turned from an asset into an unjustified financial liability for her owners.

THE UNMAKING OF END-OF-LIFE VESSELS IN ALANG

One might wonder how a small village in north-western India ended up being home to the world's largest shipbreaking industry. The answer lies partly in Alang's significant geographical advantages, which include a high tidal range, a 15° slope that makes it easier for ships to run aground, and a rocky bottom surface. These characteristics have played a significant role in its development as one of the world's largest shipbreaking yards.[44] With the beaching of a Russian cargo ship, the *M. V. Kota Tenjong*, on 13 February 1983, shipbreaking commenced here as a full-time activity (fig. 4).

An estimated 45,000 ocean-going ships currently operate on the world's seas, and an average of 700 ships are sent for demolition every year.[45] Since it began shipbreaking, Alang has beached a total of 7,891 vessels, representing 62.40 million metric tonnes of light displacement tonnage (LDT), and the industry is continuing to grow.[46] All kinds of end-of-life vessels, including large supertankers, ocean liners, crude oil tankers, ro-ro ships, animal carriers and container ships, make their final journeys to these yards. The shipbreaking industry undoubtedly remains a great source of revenue for Gujarat as it generates large quantities of re-rollable steel and accounts for 15% of the country's total steel output.[47] It thus acts as an alternative to the non-renewable resource of ore, while representing a valuable source of supply for second-hand goods. Deborah Breen

---

44 Other shipbreaking yards in India are located in Mumbai, Kolkata and Vishakhapatnam.

45 International Federation for Human Rights, Fidh: Where do the "floating dustbins" end up? Labour Rights in Shipbreaking Yards in South Asia: The cases of Chittagong (Bangladesh) and Alang (India), https://www.fidh.org/IMG/pdf/bd1112a.pdf (accessed 12.03.2019).

46 Gujarat Maritime Board: Ship Recycling Yards: No. of Ships and LDT, https://gmb ports.org/ship-recycling-yards (accessed 12.03.2019). Light Displacement Tonnage (LDT) is a measure expressed in metric tonnes and represents at best the scrap value of the ship. It is the actual weight of the ship excluding cargo, fuel, ballast water, stores, passengers and crew.

47 NL, Greenpeace: "Shipbreaking in Asia: Unregulated Trade Contributes to Concentration of Dangerous Activities in Developing Countries", 7 May 2019, Greenpeace NL collection, 1999 II, 1340 IV, Archive of Greenpeace NL, International Institute of Social History.

notes similarly that in Bangladesh's case, re-rolled steel from beached ships provided up to 80% of the steel used in the local construction industry in the last
two decades of the 20th century.[48]

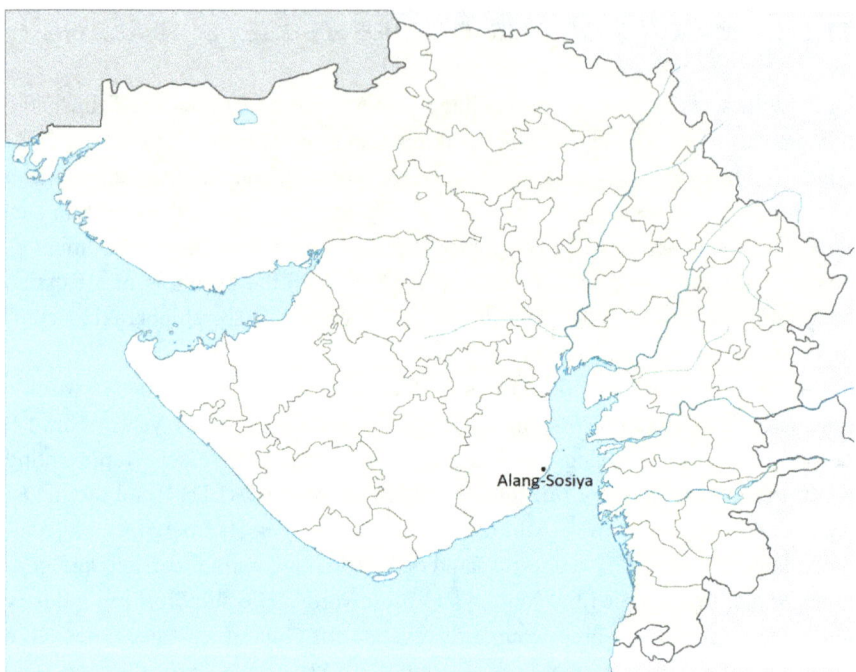

Figure 4. Location map of Alang-Sosiya in the state of Gujarat, India (Map:
https://commons.wikimedia.org/wiki/File:India_Gujarat_location_map.svg, CC BY-SA 3.0,
altered, accessed 25.09.2019).

As soon as one sets foot in the city of Bhavnagar, 50 km away from the villages
of Alang-Sosiya, a difference can be felt in the surroundings. Marine fittings,
machinery, spare parts and all kinds of knick-knacks related to seafaring vessels
can be seen hanging on both sides of the road as far as the eye can see. This visual
sight is followed by a heightened sense of a strong burning smell and the thunk
of hammers ringing in one's ears as one enters Alang. From a distance, vessels at
various stages of their dismantling processes are visible. These range from ships

---

48 Breen, Deborah: "Constellations of Mobility and the Politics of Environment: Prelim-
    inary Considerations of the Shipbreaking Industry in Bangladesh", in: Transfers 1, 3
    (2011), p. 24–43.

being stripped of their fittings to a ship whose nose is being cut auspiciously, denoting that the salvaging process has just begun (fig. 5), half-broken ships and the skeletal remains of a vessel lying along the coast awaiting a few final steps before it disappears completely.

Figure 5. Shipbreaking in progress with a cut being made on the nose of the ship (Photo: Ayushi Dhawan).

Shipbreaking is defined as the "process of dismantling a vessel's structure for scrapping or disposal whether conducted at a beach, pier, dry dock or dismantling slip".[49] It includes a wide range of activities, from removing all types of machinery and equipment to cutting down the ship's infrastructure. The industry was first developed in the USA, the UK and Japan during the Second World War, since a huge number of ships had been damaged and there was an urgent demand for steel. In the 1960s it moved to less-industrialised European countries such as Spain and Italy. In the 1970s, ship-breaking centres relocated to Asia,

---

49 Pasayat, Arijit, S. H. Kapadia: Writ Petition (C) No. 657 of 1995 (With SLP) No. 16175/1997, C. A. No. 7660/1997 and Suo Motu Con. Petition 155/2005, http://ec. europa.eu/environment/waste/ships/pdf/indian_order2007.pdf (accessed 15.03.2019).

first to Taiwan and South Korea, and then during the 1980s to China, Bangladesh, India, Pakistan, the Philippines and Vietnam.

Scholars have explained this constant shift in the centres of demolition from developed to developing countries in various studies.[50] They argue that the shipbreaking industry moves and relocates to wherever it is easiest to externalise social and environmental costs. This relocation from the Global North to the Global South has been complemented and further strengthened by rising demand for steel in developing domestic markets, lax environmental regulations and a cheap workforce.

Obsolete vessels that arrive in Alang for scrapping are primarily sold on the basis of weight to shipbreaking companies by two methods, either directly or through cash buyers.[51] There is a legislative framework in place to regulate the industry. Government departments such as the Gujarat Maritime Board (GMB), Gujarat Pollution Control Board (GPCB), Explosives Department (consulted for oil tankers), Customs Department and Atomic Energy and Research Board (consulted if there are radioactive materials on board) inspect the vessel and issue relevant certificates, and only then can the scrapping process begin to take place.[52] Once the vessel is beached and reaches the shore, the engine is shut down, anchors are dropped to the seabed and electricity on board the ship is cut off completely. All operations after the beaching process take place directly on the plots located along the coast of Alang. Current estimates suggest that there are currently around 153 plots in operation which are generally leased to shipbreakers by the state of Gujarat for ten years.[53]

---

50 Haldar, Stuti/Dutta, Indira (eds.): Alang Shipbreaking Industry: An Ecological Distribution Conflict, New Delhi: Allied Publishers Pvt. Limited 2017; Clapp, Jennifer: "The toxic waste trade with less-industrialized countries: economic linkages and political alliances", in: Third World Quarterly 15, 3 (1994), p. 505–518; Sinha, Saujanya: "Ship Scrapping and the Environment – the buck should stop!", in: Maritime Policy and Management 25, 4 (1998), p. 397–403.

51 The ship owner may sell the ship directly to a shipbreaking company by taking charge of its transportation to the final destination (in this case the shipbreaking yards), or preferably sell it through a broker. Alternatively, a ship owner may sell the ship to a "cash buyer" company such as GMS or the Wirana Shipping Company. These companies buy the ships and resell them to shipbreakers.

52 Oral Interview with Port Officer, Captain Sudhir Chadha, Gujarat Maritime Board, 31 May 2018.

53 Oral Interview with R. M. Ram Patel, Vice-President Alang Sosiya Ship Recycling and General Workers' Association, 1 Jun. 2018.

Before the dismantling process starts, a ship's fuel tanks are drained to prevent any accidental explosions on board. An army of workers then go on board the ship to strip it of its fittings, which include electronics, furniture, cooking ware, machinery, wiring, plumbing and many other items that are later sold in second-hand markets. Only after this step does the actual salvaging process start, with an auspicious cut being made on the nose of the ship using acetylene torches. Further openings are made in the hull of the ship. These serve two purposes: they allow more light into the vessel and also act as escape routes for workers in the event of accidental fire.

The ship is then cut piece by piece; the workers begin dissecting the front portion and gradually work their way towards the very end. Even the most impregnable and sturdiest ships are torn down in a matter of months by the arduous labour of workers assisted by a modest variety of tools and machines such as sledgehammers, acetylene torches, winches and cranes.[54] The time taken for the complete demolition process depends entirely on the type of vessel in question. For instance, an oil tanker takes comparatively less time to scrap than a passenger liner as the latter has a more complex inbuilt structure than the former. At Alang, an average ship of 40,000–60,000 LDT is broken in approximately three to five months.

These yards employ around 40,000–60,000 workers in total for shipbreaking operations every year. Workers from different parts of the country, especially Odisha, Uttar Pradesh, Bihar and West Bengal, migrate to Alang in search of better employment opportunities. Historian Geetanjoy Sahu states that since a majority of workers working at shipbreaking yards are migrants, no database has been created or maintained indicating the total number of workers employed in individual yards.[55] This problem is further complicated by the availability of ships for demolition at a particular yard at a given moment in time; workers tend

---

54 Kot, Michael: Shipbreakers, https://www.youtube.com/watch?v=5jdEG_ACXLw (accessed 12.03.2019); Rane, V. Prathamesh: Echoes of Shipbreaking, https://www.youtube.com/watch?v=vV3M4jqD-Sg (accessed 12.03.2019). These documentaries vividly explore shipbreaking along the beaches of Alang, showing how workers break dilapidated vessels from the Global North, live with and dispose of hazardous wastes and transform these geriatric vessels into scrap metal that is used in downstream industries, in turn benefiting the local economy of India.

55 Sahu, Geetanjoy: "Workers of Alang-Sosiya: A Survey of Working Conditions in a Ship-Breaking Yard, 1983–2013", in: Economic and Political Weekly XLIX, 50 (2014), p. 52–59.

to switch employers quite often depending on the availability of ships and requirements at the yards.[56]

Depending on their skills and experience, workers are very often categorised into *mukadams* (supervisors), gas cutters (working on the ships and at the yards), winch and crane operators, loaders and yard cleaners. They are paid daily wages according to these classifications. During the field work in summer 2018, a gas-cutter's wage was around 800 rupees ($11), compared to a yard cleaner who earned around 200 rupees ($3). In comparison to the yard workers, a rickshaw puller, for instance, would have to work more than a week or two to earn the same amount of money. The constant search for better employment opportunities than in their home states therefore brings a lot of workers to Alang. Apart from direct employment, the yards create indirect employment opportunities for tens of thousands of workers employed in downstream industries, such as re-rolling mills, oxygen plants and the real estate market, thereby contributing to the economic growth of the country.

Shipbreakers argue that 97% of a ship's contents are recyclable. Before the dismantling process, materials that are extracted from the carcass of the ship, like ferrous objects, non-ferrous objects, wood, glass, plastic, machinery and other equipment, are neatly separated out and sold in second-hand markets. Larger sheets of steel are sent to re-rolling mills where they are converted into rods and bars and then supplied to local construction industries. Other materials such as loose asbestos, metallic waste, plastic scrap and broken glass, which are often deemed as having no commercial value and are categorised as residual wastes, end up in a state-owned secured landfill site in Ahmedabad.

Since its inception, however, the industry has been frequently criticised by environmental activists and NGOs for scrapping vessels through the beaching method. They argue that shipbreaking is a hazardous activity as it exposes workers and the environment to chemicals that are released during the demolition process which takes place on beaches.[57]

---

56 Oral Interview with R. M. Ram Patel, Vice-President Alang Sosiya Ship Recycling and General Workers' Association, 1 Jun. 2018.

57 NL, Greenpeace: "Shipbreaking in Asia: Unregulated Trade Contributes to Concentration of Dangerous Activities in Developing Countries", 7 May 2019, Greenpeace NL collection, 1999 II, 1340 IV, Archive of Greenpeace NL, International Institute of Social History.

SECOND-HAND MARKETS: NEW USES OF
DECOMMISSIONED SHIPS

In hindsight, for the shipbreaking company in Alang, the *SS Blue Lady* was a sought-after source of ferrous and non-ferrous scrap, precisely 47,689.10 metric tonnes of LDT.[58] For traders at second-hand markets, the ship's body was a rusting carcass that contained traditional and contemporary fittings. The shipbreakers' bills of sale and purchase reveal that the ship contained 10 passenger decks, 5 diesel generators, 6 steam generators, 2 diesel-driven emergency generators, 2,000 televisions, 800 refrigerators, 40 computers, 1,500 wall clocks, 120 vacuum cleaners, 1,200 mirrors, 1,500 mattresses, 2,000 pillows, 600 navigational charts, 850 sets of cutlery, 200 plastic buckets and many more items.[59]

Even as the *SS Blue Lady* was broken piece by piece and rivet by rivet in Alang, her steel was being used in the local construction industries and her artefacts and fittings had been recovered through auctions and were now either in use by new owners or even preserved in public and private collections across the world. For instance, the prow of the ship has been proudly installed in the port of Le Havre. The city authorities purchased it at auction for €150,000. As the prow was being installed, Le Havre's Mayor Luc Lemmonier stated, "[t]o have in our city a vestige of this symbolic liner, a piece of our heritage, is highly symbolic".[60]

Every year, as so many vessels are scrapped in Alang, second-hand markets are lined up on either side of the road to the yards in Alang, stretching for approximately 10 km. Every reusable part found in a ship, ranging from consumer goods such as furniture, crockery and carpets to kitchen equipment such as toasters, ovens and sinks, machinery like engines, generators and compressors, and life-saving equipment, are found at these markets at very cheap prices (fig 6). Traders dealing in various objects visit the vessel, once the shipbreaker has received appropriate clearances from the regulatory authorities. They negotiate with the shipbreakers for the entire cache of goods in their categories. These

---

58 Blue Lady, SS: Bills of Sale and Purchase, 3 Jun. 2018, Priya Blue Industries Pvt. Ltd collection, 2006, Office of Priya Blue Industries Pvt. Ltd, Bhavnagar, Gujarat.

59 Ibid.

60 Bond, Mary: "Tip of the bow of the former France back in Le Havre", in: Seatrade Cruise News, 27 Sep. 2018, p. 1; Normandy, France 3: The Liner France again visible in Le Havre, https://www.youtube.com/watch?v=ptXFig3yniA (accessed 12.05.2019); Normandy France 3: The Incredible story of the nose of France, https://www.youtube.com/watch?v=N4kBJQJA7ZM (accessed 12.05.2019).

markets demonstrate how products that are rejected by the Western world are re-integrated into the local economy, exposing the other side of globalisation. Geographers Mike Crang and Nicky Gregson, who have investigated shipbreaking practices in Bangladesh, have noted similar practices in Chittagong's second-hand markets. As they point out, "the proud boast of the industry is that 99 percent of an end-of-life ship is recycled".[61]

Figure 6. Second-hand lifeboats along with reusable metal springs in a local market in Alang (Photo: Ayushi Dhawan).

This burgeoning trade in recycled goods attracts hoteliers, factory owners from various industrial centres such as Delhi, Punjab, Chennai and Hyderabad, art collectors, homemakers and ship enthusiasts, who come looking for the remains of vessels. Most products can be found at one-fifth of the new price. Since machinery, iron and steel products are sold on the basis of weight, buyers can enjoy a

61 Crang, Mike/Gregson, Nicky/Ahamed, Farid/Ferdous, Raihana/Akhter, Nasreen: "Death, the Phoenix and Pandora: transforming things and values in Bangladesh", in: Alexander, Catherine/Reno, Joshua (eds.): Economies of Recycling: The Global Transformation of Materials, Values and Social Relations, London/New York: Zed Books 2012, p. 59–97, here p. 65.

profitable deal, and at times they come not just from the local areas but also from overseas. In terms of quality, traders vouch for their products, constantly reiterating that these products are often not easily available in India and that they are of good quality as Western shipbuilders use top-notch equipment to minimise repairs during the product's operational life.[62]

CONCLUSIONS

Through maintenance and repair, the operational life of objects can be increased. But as the long and eventful voyage of the *SS France* reveals, sometimes prohibitively expensive repairs and maintenance costs are avoided by owners and the objects are disposed of and sent for recycling. The illegal export of the *SS France* from the Global North to the Global South for scrapping, as she plied the seas with different names on behalf of a string of owners, demonstrates that the production, repair, maintenance, reuse and recycling of technical artefacts are therefore directly interwoven with questions of waste and disposal. The journey of the *SS France* is emblematic of many other end-of-life vessels that end up at the shipyards in Alang, where distinct dichotomies of new v. old, functional v. non-functional, valuable v. worthless all become blurred as vessels with both illustrious and uneventful histories are broken alike. They are reverted back to tonnes of re-rollable steel and streams of reusable, toxic and non-reusable materials that are reintegrated and remobilised into the local economy or landfilled at Ahmedabad by the arduous labour of unmaking things, a process that remains toxic and life-giving at the same time.

---

62 Oral interviews with traders at second-hand markets: Sara Enterprises (dealing in generators and marine machinery), Solas Marine (dealing in spare machinery), Bhagvati Traders (dealing in clothes, fans and other retail items), Jodiyar Traders (dealing in ropes, nets, pumps and ship machinery) on 10 Jun. 2018.

Authors

Dhawan, Ayushi, is currently pursuing a PhD in Environmental Humanities at the Rachel Carson Center for Environment and Society, Ludwig Maximilian University of Munich. She is a member of the DFG Emmy-Noether Research Group "Hazardous Travels. Ghost Acres and the Global Waste Economy". With a background in history, she is currently focusing on the Alang Shipbreaking Industry in Gujarat from 1983 to 2012. She has written various blog posts for Seeing the Woods, Kunsthal Extra City #Cahier 5 and Environmental History Now.

Hadlaw, Jan, is an Associate Professor in the School of the Arts, Media, Performance & Design at York University, Canada. She is currently completing a cultural and material history of the modern American telephone. Recent publications: "Design Nationalism, Technological Pragmatism, and the Performance of Canadian-ness: The Case of the Contempra Telephone", in: Journal of Design History 32, 3 (2019), p. 240–262; "'Mysteries of the New Phone Explained': Introducing Dial Telephones and Automatic Service to Bell Canada Subscribers in the 1920s", in: Imhotep-Jones, Edward/Adcock, Tina (eds.): Made Modern: Science and Technology in Canadian History, Vancouver: University of British Columbia Press 2018, p. 143–165.

Krebs, Stefan, is an Assistant Professor for Contemporary History and Head of Public History at the Luxembourg Centre for Contemporary and Digital History (C²DH). He is currently working on the history of maintenance and repair in Luxembourg and the industrial past of the "Minett" region. Recent publications: Krebs, Stefan/Schabacher, Gabriele/Weber, Heike (eds.): Kulturen des Reparierens: Dinge – Wissen – Praktiken, Bielefeld: transcript 2018 (open access); Krebs, Stefan: "Testing Spatial Hearing and the Development of Kunstkopf Technology, 1960–1981", in: Tkaczyk, Viktoria/Mills, Mara/Hui, Alexan-

288 | The Persistence of Technology

dra (eds.): Testing Hearing: The Making of Modern Aurality, Oxford/New York: Oxford University Press 2020, p. 213–242.

Lean, Thomas, is a Research Associate in the History Department of the University of York. Prior to this he was a project interviewer for An Oral History of the Electricity Supply Industry and An Oral History of British Science, for the British Library's "National Life Stories" initiative. He is currently working on a history of pesticides in agriculture. Recent publications: Electronic Dreams: How 1980s Britain Learned to Love the Computer, Bloomsbury: Sigma 2016; "The Life Electric: Oral History and Composure in the Electricity Supply Industry", in: Oral History 46, 1 (2018), p. 55–66.

Łotysz, Sławomir, is a Professor at the Institute for the History of Science at the Polish Academy of Sciences in Warsaw, Poland. Recent publications: "Knowledge as Aid: Locals, Experts, International Health Organizations and Building the First Czechoslovak Penicillin Factory, 1944–49", in: Reinisch, Jessica/Brydan, David (eds.): Internationalists in European History: Rethinking the Twentieth Century, London: Bloomsbury 2021, p. 140–157. He is currently working on a monograph about the environmental history of the Pripet Marshes.

Lucsko, David N., is a Professor and Chair of the Department of History at Auburn University. He is the author of: The Business of Speed: The Hot Rod Industry in America, 1915-1990, Baltimore: Johns Hopkins University Press 2008 and: Junkyards, Gearheads, and Rust: Salvaging the Automotive Past, Baltimore: Johns Hopkins University Press 2016. He is currently working on a monograph about old-car restoration.

Marhold, Karsten, is a Research Fellow at the Fund for Scientific Research (F.R.S.-FNRS) and a member of the Modern & Contemporary Worlds research centre at the Université Libre de Bruxelles (ULB). The chapter in this book is drawn from his doctoral research, which examines the development of electric vehicles in the 1970s from an STS perspective.

Petrova, Mariya, is a doctoral researcher at the Leibniz Institute for Regional Geography in Leipzig, Germany, working on public transportation infrastructures and policies in post-Soviet Uzbekistan. Prior to this, she conducted research on the history of building and housing in Soviet Samarkand, Uzbekistan, as a member of the project "A Global History of Technology, 1850–2000" at the Technical University of Darmstadt. Recent publications: Nah am Boden. Privater

Hausbau zwischen Wohnungsnot und Landkonflikt im Samarkand der 50er- und 60er-Jahre, Berlin/Boston: De Gruyter 2021 (open access).

Tan, Ying Jia, is an Assistant Professor of History and East Asian Studies at Wesleyan University in Middletown, CT. He is the author of: Recharging China in War and Revolution, 1882–1955 (2021), which is published under the Sustainable History Monograph Pilot, an initiative supported by funding from the Andrew W. Mellon Foundation, making its e-book edition available as an open access volume from Cornell Open and other repositories.

Van der Straeten, Jonas, works as a postdoctoral researcher at the Technical University of Darmstadt on the project "A Global History of Technology, 1850–2000", which is funded by the European Research Council. His current research focuses on the temporality of technology, especially in the region of Central Asia. Recent publications: van der Straeten, Jonas: "Borderlands of Modernity. Explorations into the History of Technology in Central Asia, 1850–2000", in: Technology and Culture 60 (2019), p. 659–687; van der Straeten, Jonas/Hasenöhrl, Ute: "Connecting the Empire: New Research Perspectives on Infrastructures and the Environment in the (Post)Colonial World", in: NTM Journal of the History of Science, Technology and Medicine 24 (2016), p. 355–391.

Weber, Heike, is Professor of History of Technology at the Technische Universität Berlin. Her main research focus lies at the intersection of consumption history, environmental history and history of technology. She has worked on everyday 20th-century technologies and the interlinked history of media and mobility, and she is currently studying the history of waste, recycling and repair, thereby pushing the history of technology beyond its traditional focus on production, consumption and use towards issues of obsolescence, decay and disposal. Recent publications: Heßler, Martina/Weber, Heike (eds.): Provokationen der Technikgeschichte. Zum Reflexionszwang historischer Forschung, Paderborn: Schöningh 2019; Krebs, Stefan/Schabacher, Gabriele/Weber, Heike (eds.): Kulturen des Reparierens: Dinge – Wissen – Praktiken, Bielefeld: transcript 2018 (open access).

Historical Sciences

Federico Buccellati, Sebastian Hageneuer,
Sylva van der Heyden, Felix Levenson (eds.)
Size Matters –
Understanding Monumentality
Across Ancient Civilizations

2019, 350 p., pb., col. ill.
44,99 € (DE), 978-3-8376-4538-5
E-Book: available as free open access publication
PDF: ISBN 978-3-8394-4538-9

Sebastian Haumann, Martin Knoll, Detlev Mares (eds.)
Concepts of Urban-Environmental History

2020, 294 p., pb., ill.
29,99 € (DE), 978-3-8376-4375-6
E-Book:
PDF: 26,99 € (DE), ISBN 978-3-8394-4375-0

Jesús Muñoz Morcillo, Caroline Y. Robertson-von Trotha (eds.)
Genealogy of Popular Science
From Ancient Ecphrasis to Virtual Reality

2020, 586 p., pb., col. ill.
49,00 € (DE), 978-3-8376-4835-5
E-Book:
PDF: 48,99 € (DE), ISBN 978-3-8394-4835-9

All print, e-book and open access versions of the titles in our list
are available in our online shop www.transcript-verlag.de/en!

Historical Sciences

Monika Ankele, Benoît Majerus (eds.)
Material Cultures of Psychiatry

2020, 416 p., pb., col. ill.
40,00 € (DE), 978-3-8376-4788-4
E-Book: available as free open access publication
PDF: ISBN 978-3-8394-4788-8

Judith Mengler, Kristina Müller-Bongard (eds.)
Doing Cultural History
Insights, Innovations, Impulses

2018, 198 p., pb., col. ill.
34,99 € (DE), 978-3-8376-4535-4
E-Book:
PDF: 34,99 € (DE), ISBN 978-3-8394-4535-8

Jochen Althoff, Dominik Berrens, Tanja Pommerening (eds.)
Finding, Inheriting or Borrowing?
The Construction and Transfer of Knowledge
in Antiquity and the Middle Ages

2019, 408 p., pb., ill.
54,99 € (DE), 978-3-8376-4236-0
E-Book: available as free open access publication
PDF: ISBN 978-3-8394-4236-4

**All print, e-book and open access versions of the titles in our list
are available in our online shop www.transcript-verlag.de/en!**

GPSR Authorized Representative: Easy Access System Europe, Mustamäe tee 50, 10621 Tallinn, Estonia, gpsr.requests@easproject.com

www.ingramcontent.com/pod-product-compliance
Lightning Source LLC
Chambersburg PA
CBHW061603120626
46550CB00004B/1602